T0076040

Stories, Dice,
and Rocks
That Think

Also by Byron Reese

Wasted
The Fourth Age
Infinite Progress

Stories, Dice, *and* Rocks That Think

How Humans Learned to See the Future— and Shape It

BYRON REESE

BenBella Books, Inc.
Dallas, TX

BenBella Books, Inc.
10440 N. Central Expressway
Suite 800
Dallas, TX 75231
benbellabooks.com
Send feedback to feedback@benbellabooks.com

BenBella is a federally registered trademark.

Printed in the United States of America
10 9 8 7 6 5 4 3 2 1

Library of Congress Control Number: 2022935037
ISBN 9781637741344 (hardcover)
ISBN 9781637741351 (electronic)

Editing by Alexa Stevenson and Gregory Brown
Copyediting by Scott Calamar
Proofreading by Lisa Story and Ashley Casteel
Indexing by WordCo
Text design and composition by PerfecType
Cover design by Pete Garceau
Cover image © iStock / GuidoVrola
Printed by Lake Book Manufacturing

To Sarah, Michael, John & Peter

CONTENTS

ACT II | DICE

ACT III | ROCKS THAT THINK

INTRODUCTION

For the past couple of centuries, our species has been playing an existential game of Mad Libs, trying to fill in the blank on what seems like a pretty straightforward sentence: "Humans are the only creatures that _____." We've tried "makes tools," "uses language," "is conscious," "controls fire," "has culture," "reasons," and (per Mark Twain) "blushes, or needs to." But each time a new answer to the question of what makes us unique is offered, it is immediately pounced on by naysayers eager to disprove it and to show that there really isn't anything that special about us at all—that we are just another animal.

But common sense tells us that simply isn't true. Clearly, we are radically different from the other creatures on this planet. Who can deny this? We are Earth's preeminent life-form by such a wide margin that there isn't even a distant second. Look around. Where are the Bronze Age beavers? The Iron Age iguanas? Or the preindustrial prairie dogs? Have you ever met a penguin poet? Pheasant philosopher? Or platypus playwright? No, because they don't exist. Our planet is populated by just two types of creatures: us and a giant menagerie of beings so unlike us that the tiniest overlap is cause for curious wonder.

I'm not down on animals. Without a doubt, they feel pain as we do; their suffering is as real as ours, and because they can suffer, they should be given legal protections. Further, I'm not arguing that they don't have emotions or even that they are not conscious. Perhaps they are. I'm solely interested in *animal cognition*, which I do believe is entirely different from human cognition. In that difference lies the answer to the perplexing question posed above: If humans and animals are the same *sort* of thing, differing only by degrees, why is our outcome so dramatically different from theirs?

Obviously, it's not our bodies that give us preeminence. We have animal bodies. Good ones, to be sure, with unusually long life spans and an amazing ability to repair themselves, but that isn't what distinguishes us. It's our minds. Something about them makes us so different from animals that we are almost aliens by comparison. I think it's this: We are endowed with a temporal mental plasticity that enables our minds to roam freely through time, untethered from the here and now. Our thoughts flow effortlessly from the present to the past to the future. We can remember what happened yesterday and use it to speculate on what might happen tomorrow; we can recall our childhood and contemplate our old age. We can imagine many different futures, predict what will likely happen, and try to exercise control over it. We are the architects of our tomorrows, the shapers of our destinies. No other creature on Earth even knows that there is a future, or a past for that matter; instinctual behavior aside, animals live outside of time. But this knowledge of ours comes at a price, for it reveals our mortality. As essayist Jorge Luis Borges put it, "Except for man, all creatures are immortal, for they are ignorant of death."

There was a time when creatures that looked like us *were* animals, and they, too, didn't know there was a future or a past. How did we get from there to a point where we could think about the future; influence it; and, finally, perhaps master it? This book tells the story, in three acts, of how our species learned to escape the perpetual present. Act I is how we developed the cognitive ability to mentally time travel. It starts far in the distant past, millennia ago, and explores how we acquired language *as a mental construct*, which gave us the capacity for thought, which we only later began externalizing in a spoken form in order to communicate with others. That mental language became the voice in our head, one that we used to imagine stories about possible immediate futures—running different scenarios in our minds of the ways that events could unfold. Later, we started externalizing those as well, telling stories to each other. We were story*thinkers* before we became storytellers.

The thing that transformed us—the radioactive spider that bit us—was something so rare and so serendipitous that it evidently has never happened to any other creature. That's why there are no Bronze Age beavers. When we got our mental superpowers, fifty thousand or so years ago, we became fully "us," with our language, art, music, and all the rest. With this range of new abilities, we were able to draw upon the past to imagine multiple futures and predict which of them would happen. This gave us mastery of the planet in an evolutionary blink of an eye. With it, we invented agriculture, created cities, devised writing systems, divided into nations, and explored the world.

But we wanted more. We wanted to *systematize* prediction, turning it from an art to a science. We accomplished this. But doing so required a new understanding of the nature of reality, of why the future unfolds the way it does. That's the story told in Act II, which begins in 1654 in France, when two mathematicians trading correspondence invented what we now call probability theory. With it, we had a science for seeing into the future, and we used it to build the modern world. It became the cornerstone of a dozen sciences, from mundane meteorology to exotic quantum physics. Sociologists used it to create demography, while biologists used it to pioneer medical research. It became the basis of our financial system, the insurance industry, the capital markets, and, well, the entire world economy. All commerce was based on it, on using probability to predict everything from inventory levels to consumer demand. Virtually all of the modern world, in all its complexity, sits atop that science of seeing the future.

Three centuries passed between creating the science of probability and hitting a biological limit of what our intellects, as amazing as they are, could accomplish with it. So we began building machines that could employ the science far better than we could. The curtain closes on Act II in 1954, as we booted up the first all-transistor computer.

Act III opens on that world, where events transpired quickly. Transistor-based computers rapidly grew in capability, and we invented a science called artificial intelligence, whose explicit goal was to teach the new machines how to think as we did. The hope was that with their lightning speed, they could solve probability problems that were beyond our capabilities, allowing us to predict the future ever more accurately. At the same time, we began attaching electronic sensors to the computers, allowing them to see and hear the world on their own and to interact with it. They would gather near-infinite amounts of data, and the belief was—and still is—that all that data, combined with near-unlimited processing power, will give us the ability to see the future as accurately as we see the present. If we can do this, we will become true masters of fate.

We are living at the dawn of Act III, but perhaps we can already see into the future well enough to confidently predict how this act will turn out. But I don't want to spoil the ending in the introduction, so let's take Lewis Carroll's advice and begin at the beginning. Pour yourself a drink, sit back, and get comfortable, because I have quite a story to tell you.

Act I

Stories

Two Million Years Ago

How did we come to think about the future? To be able to recall the past and plan ahead? How did we start thinking in language and then vocalizing what we thought? How did we develop mental stories that enabled us to imagine multiple different futures?

The answer to these questions is the tale of how we came to be something very different from the other creatures on this planet. No matter how highly you regard apes, dolphins, honeybees, or ants, there is no doubt that our outcome as a species is vastly beyond theirs. Consider all humanity's accomplishments: the monuments we've built, the great art we've created, the cities we've erected, the technology we've fashioned. Look at our legal codes, our systems of government, our scientific achievements, and the exploits of our greatest heroes. What other species has a millionth of that?

Take the beaver as an example. Beavers build pretty good dams, but they are the same dams that beavers have been building since time immemorial. They haven't added hydroelectric capabilities or started using cement. The beaver doesn't even know *why* it's building a dam. If you place a recording of the sound of running water in a field—nowhere near any water—and a beaver happens by, it will instinctively build a dam over the device.

To see what made us so special, our story needs to begin two million years ago, long before humans walked the earth. Over the next few chapters, we will travel from that time to just about fifty thousand years ago, when creatures just like you and me appeared on the scene, seemingly out of nowhere—creatures with thought, language, stories, art, music, culture, and technology.

Homo Erectus

From about two million years ago to about a hundred thousand years ago, a species called *Homo erectus* (upright man) roamed the world, all over Africa, Europe, and Asia. By any measurement, *erectus* was a successful species, enduring and prospering for two million years, ten times longer than our own species thus far. But longevity and success don't necessarily correlate with intelligence and ability. You can't get much dumber than bacteria, and they'll outlive us all. This is my polite attempt to say I don't think *erectus* was a particularly smart creature. It probably would be regarded today as a particularly talented type of ape.

But *erectus* was a tool user, the creator of something we call an Acheulean (uh-SHOO-lee-un) hand ax. It looks like a large arrowhead in the shape of a teardrop. We see them appear in the fossil record when *erectus* came on the scene, and they disappeared when *erectus* exited. These tools were used for two million years, and hundreds of thousands of them have been found across three continents. There are so many that you can buy one on eBay—an actual tool used a million years ago—for a hundred bucks.

Doesn't that mean that *erectus* had at least a bit going on upstairs? No. In fact, I think these axes are the open-and-shut case against *erectus*'s mental ability, for while they show some improvement over time, it isn't all that much. Two of them that were made a million years apart could be placed side by side on a table, and you wouldn't be able to tell which was newer. Even the experts date examples to periods spanning five hundred thousand years. How is this possible? How can you have a tool that gets used for something like *eighty thousand generations* but never gets appreciably any better? Let me say that again for emphasis. Eighty. Thousand. Generations. With no change. In all that time, no one looked at their hand ax and thought, "You know, this would work better if . . ." They sound like beavers who just built the same basic dam throughout the ages, not like us.

In fact, in a paper called "The Acheulean Handaxe: More Like a Bird's Song Than a Beatles' Tune?" the authors make just such a case, exploring the idea that the hand ax wasn't even a cultural object but a genetic one. They point out that the lack of change over so much time would have been impossible if the *knowledge* of how to make the ax had been passed down. Think of the telephone game, when one person whispers something to the next person who is supposed to repeat it to the next one. After a dozen or so people, the message comes out garbled. If an *erectus* simply tried to copy his dad's ax, it would have been a bit different. After eighty thousand generations, it inevitably would have drifted even without innovation. Each copy should have varied from the last by about 3 percent, it is estimated. And over time, this copy of a copy of a copy of a copy would gradually be increasingly different. Further, the similarity between axes found thousands of miles apart over hundreds of thousands of years suggests *erectus* may have had a solely innate ability to make this object. Even in areas where the prey, climate, or terrain was different, they still made the same tool, much like the way a bird will build the same nest—the only one it knows how to make. So even the one piece of technology we give them credit for may have been something they didn't even invent, any more than a bird "invents" how to build its distinctive nest. They wouldn't have even understood why they were making them any more than the beaver understands the dam it is building over the tape recorder playing the sound of running water. I know that's hard to wrap our minds around, but that's because we have actual minds, which *erectus* lacked.

More than a few scholars would take issue with my assessment of the mental ability of *erectus*. I grant that. But think about it for a minute. Using twenty years as a proxy for a generation, then the time between Kitty Hawk and the moon landing was just three generations. Between the invention of the telephone and the iPhone, seven generations passed. The time between the first coin and today's vast, interconnected financial system was just 125 generations. Finally, the time between the first written word etched into a piece of soft clay and William Shakespeare was just 250 generations. How does anything that is remotely like us go eighty thousand generations without noticeable technological advance? They don't, because they were nothing like us.

I know it sounds like I have an ax to grind with *erectus*, as if I was bullied by one as a child. The reason I am stressing this is to draw a sharp distinction between us and them, the supposed protohumans. They did not gradually morph into us after some slow but steady advance over eons. Definitely not. There are eighty thousand good reasons to believe this. If they had gradually turned into us, we would see it in their tools.

No, what happened that turned us into us happened in the blink of an eye, not over the eons. As we'll soon see, one day there was nothing like us in the archaeological record; then, in an instant, there we were, fully formed, fully us. The change was so dramatic, in fact, that it may have just happened to one person one time in history, on a certain date and at a certain time. We might all be descendants of that one fortuitous individual. That's our next stop.

Becoming Us

Fifty thousand years ago (YA)—or perhaps a few thousand years earlier—the archaeological record suggests that something dramatic and unprecedented happened to the human race. Something completely transformative that gave us, in the blink of an eye, technology, art, music, imagination, creativity, language, stories, and a notion of the future. This chapter will lay out the facts of what we know and the questions it raises, and the next chapter will try to make sense of it all.

If we pass judgment on *Homo erectus*'s intellect for his lack of technical advancement over nearly two million years, what exactly are we looking for in the archaeological record that would recognizably be "us"? Evidence of a sophisticated spoken language or storytelling tradition would be great, but spoken words leave no fossils. Before the invention of audio recording, all sound was fleeting, existing in an instant and then vanishing without a trace. Italian inventor Guglielmo Marconi believed that sound waves are never completely gone and that with sensitive enough equipment you could hear, as was his wish, the Sermon on the Mount. But since we lack such a device, our best proxies for knowing when we came on the scene would be a rapid technological advance along with the emergence of representative art—that is, art that depicts something, as opposed to simply geometrical figures. If you found a carving of some fantastic imaginary creature or a painting of a hunting scene, you could be confident that it meant something to the people who made it. There was a story there. Those would be people who spoke, who had imagination. If you found such art executed with complex technology, suddenly you might recognize a kindred spirit.

Before about 50,000 YA, the evidence for such art is nonexistent. After 50,000 YA, we see it in great abundance in the form of cave art, which has been discovered on every continent except Antarctica. Western Europe especially teems with it, with hundreds of spectacularly decorated caves dating all throughout what we now call the Upper Paleolithic, from 40,000 YA to 12,000 YA.*

How would you suppose this art developed and matured? One might assume that the art from 40,000 YA would consist of stick figures, and that over time a bit more sophistication was gradually rolled in. But this isn't the case at all, for it seems to have appeared fully formed in all its beauty from the beginning. As Werner Herzog, who filmed a documentary on the Chauvet Cave in southern France, described it, "It's not that we have what people might call the primitive beginnings of painting and art. It is right there as if it had burst on the scene fully accomplished. That is the astonishing thing, to understand that the modern human soul somehow awakened." Art critic John Berger, who compared the artists of Chauvet favorably with those of the Renaissance, seemed to agree, writing, "Apparently art did not begin clumsily . . . There was a grace from the start. This is the mystery, isn't it?" Indeed it is, as we shall soon see.

Let's explore the Chauvet Cave. It was discovered in 1994 by its namesake, speleologist Jean-Marie Chauvet,† and was found to contain prehistoric paintings of unsurpassed beauty. By this, I don't mean, "Wow, those

---

* After 12,000 YA, when the most recent ice age ended, we entered into the Neolithic era, when we adopted agriculture, domesticated animals, built permanent cities, spread to the Americas, and generally settled down to the pattern of life we enjoy today. Cave art, for unknown reasons, lost its significance, and the ancient caves were forgotten, waiting to be rediscovered.

† Speleology is the scientific study of caves. Spelunking is the recreational exploration of them.

are some nice paintings to be from such primitive humans." No, I mean that by any standard of today, they are stunningly beautiful. And what a menagerie! Over four hundred individual animals are shown, including cave bears, mammoths, lions, rhinos, and horses. And they are rendered not as copies of the same basic pattern but as individual animals with unique, for lack of a better word, personalities. They date from around 40,000 YA, making them among the oldest cave paintings we know of.

The artistry is masterful. The contours of the cave walls themselves are worked into the anatomy of the animals. They were painted on walls that had been carefully prepared by scraping. Some animals were depicted with eight legs, four of which are less distinct, so that flickering torchlight would give the illusion that the animal was running.

The great thing about Chauvet is that it was so pristine, thanks to an avalanche that closed off the entrance, sealing it up like Tutankhamun's tomb. So undisturbed are its contents that you can still see footprints in the dirt on the floor of the cave. As one of the discoverers later wrote, "Everything was so beautiful, so fresh. It was as if time had been abolished."

At roughly the same time that artisans were busy painting at Chauvet, seven thousand miles away on the island of Borneo, other artisans were painting, in a similar style, on the cave walls in Lubang Jeriji Saléh, images of a type of cattle that still roams Borneo today. A few hundred miles from there, across the eighty-mile Makassar Strait on the Indonesian island of Sulawesi, caves containing art depicting a large, piglike creature are confirmed to be five thousand years older than Chauvet.

No one is entirely sure what to make of the concurrence of all of this art in places so disparate. In the paper "Palaeolithic cave art in Borneo" describing the find at Lubang Jeriji Saléh, the authors write that "similar cave art traditions appear to arise near-contemporaneously in the extreme west and extreme east of Eurasia. Whether this is a coincidence, the result of cultural convergence in widely separated regions, large-scale migrations of a distinct Eurasian population or another cause remains unknown."

We can be confident that more caves with art will be discovered, because archaeology is, practically speaking, a new field. Before World War II, there were barely more than a hundred archaeologists in the world. Today there are thousands.

Upper Paleolithic cave art is a mysterious thing. Most significant is that the paintings are often not near cave entrances, but rather are deep into the cave—up to a mile—and difficult to reach. Frequently art is found in caves in which people never lived, so cave art wasn't a decoration painted in a dwelling. Also puzzling are the objects *not* painted on cave walls, such as trees, lakes, clouds, the sun, fire, or any of the other things you would imagine would be meaningful to those humans living in that twenty-eight-thousand-year span of history. Most tantalizing is that humans are virtually never depicted. Overwhelmingly, the artists painted animals, with the mix of species varying from cave to cave. Sometimes the animals are prey animals, while other caves will exclusively feature predators. Paintings in the caves of Lascaux in Eastern France feature more than six hundred different figures of animals, half of which are horses, but not a single reindeer, which would have been the artists' main source of meat. In Lascaux, horses are also featured more than any other creature. Paul Pettitt, professor of Paleolithic archaeology at Durham University, speculates why this is: "It's probable, although hard to demonstrate, that the horse had some totemic and/or cosmological importance," adding that horses fall "out of the rock in places, as if it's a place of birthing."

Aside from animals, two other things make up virtually all the cave paintings of this period. The first of these are abstract symbols. There are relatively few different symbols that are used in all the caves we know about. We cannot decipher what they mean and are thus left to wonder. The second are handprints, often negative handprints, created by blowing pigment through a hollow bird bone onto a person's hand pressed against the cave wall. When the hand is pulled away, a silhouette remains.* Handprints are found everywhere, from Australia to Asia to Europe and even to the

---

* Because of this technique, scientists are hopeful they can extract DNA from the residual saliva in the pigment, which would tell us a good deal more about who made them.

ants basically invented agriculture millions of years before we did, which is a little humbling. When it rains, the leafcutters seal up their nests; however, if they do this too long, carbon dioxide builds up in the nest. While they can tolerate it, the fungus can't, so the leafcutters have to be careful to open the door every now and then to let in fresh air, lest they kill their crop. Of course, no leafcutter knows about carbon dioxide, symbiosis, or anything for that matter, and yet the nest accomplishes all of this. Across the Atlantic in Africa are termites that build giant nests with elaborate ventilation systems. They deal with the rain problem by digging out giant flood chambers where rain is routed by gravity. The individual termite's understanding of convection, hydrology, and Newtonian physics is nonexistent, yet the superorganism has a high degree of proficiency in all of them, arguably on par with a human engineer.

You are a kind of superorganism. All of the cells in your body come together to make you, but none of them has a sense of humor, yet you do. There are those who believe consciousness is an emergent property of complex superorganisms. Are your cells conscious? Probably not. But you certainly are. Bees probably aren't conscious either, but is the hive? Maybe. There was a widely practiced ancient tradition that is still practiced today called "telling the bees," wherein a beekeeper informs the hive of significant events in his or her family, especially marriages and deaths. It begins with knocking on each hive one by one and relaying the news. If it is a death, then the hives are draped in black, joining the family in mourning. When the beekeeper dies, someone else has to go tell the bees.

Now the question you have probably seen coming: Is a tribe of humans a superorganism? Yes, if they have spoken language. Superorganisms are characterized by a high degree of peer-to-peer communication. In the insect world, this is often done through touch and chemical trails. The density of that information exchange gives the superorganism its emergent properties. Your brain has a hundred billion neurons packed in a small area communicating with each other at an amazingly fast rate. Your mind is believed to emerge from that activity. If the neurons were scattered around the earth and could communicate only through mail, the magic would likely be lost.

Americas. Often, they are those of men, women, and children, including very young children, and the number of left hands depicted is high, suggesting that most, but not all, of these folks were right-handed. There are also often prints from multiple hands that appear to be missing all or part of a finger. Was Paleolithic living that hard on digits? Or, as some have suggested, are these messages in sign language, not from people missing fingers? Handprints, especially of the reverse variety, seem an odd thing to be so pervasive across such a vast span of territory and time, and the fact that many walls of these prints scattered around the world all look remarkably the same only adds to the mystery, as does the fact that many caves have handprints as the only form of adornment.

Another puzzling aspect of cave art is the vastly different ages of paintings in a single cave. Chauvet has paintings dating five thousand years apart, and Altamira, located in Spain, has paintings that span more than fifteen thousand years. What caused these early humans to return to the same spots, buried deep inside mountains, for hundreds of generations? It seems an almost primordial act, the way Pacific salmon are compelled to return to where they were born to propagate.

Finally, the techniques used by the people of the Upper Paleolithic are remarkably sophisticated. They needed to be, for the ambition of the artists was high. This is an important point considering the technology that was deployed as well as the advanced planning required. The paintings they produced used shading to create depth, and they depicted fine detail such as muscle tone. The eyes of various animals are often clearly looking at different parts of the menagerie. The artisans at Lascaux used common pigments such as chalk, charcoal, and colorful clays, but they also used an uncommon mineral for deep black called hausmannite. This is striking because the closest source of that mineral is more than 150 miles from Lascaux. To make it useful as a pigment, it had to be extracted from ore after being burned in an inferno of more than 1,600 degrees Fahrenheit. Think about that. They had an ample source for black—charcoal—but it wasn't black enough for them, so they went to all this trouble to make a deeper black. Extenders, such as talc, were added to paint to bulk it up, while binders such as animal fat were added to it to make it adhere better to the walls. Furthermore, many caves have art on ceilings far above a human's reach, so there must have been scaffolding of some kind built to access those places. Often the artists sketched a broad outline on the wall first to get the composition down before they started painting.

So what does all this art mean? We can never really know, but one cannot help but speculate. Some believe it is something akin to religious texts, functioning the same way stained-glass windows taught illiterate medieval Christians the central stories of their faith. Others regard it as apologies to the animals of the hunt, or invocations to have a successful hunt, or initiation rites for boys, or rituals to promote fertility, or an attempt to absorb the power of the animals, or a form of communal activity to foster group collaboration, like a Paleolithic Amish barn raising. Others have speculated that the paintings are representations of drug-induced visions, the sorts one might have in utter darkness while tripping on something. Other theories are that they contained information that future generations would need to survive, or even that the pairings of different animals record alliances between different clans, like a truce between the prehistoric Jets and Sharks. Attempts to interpret the art sometimes take on a vibe similar to that of reading tea leaves. Meticulous studies have logged the exact locations of different animals in caves, in the belief that there may be an underlying pattern that can be discovered and interpreted.

In France, where many magnificent caves have been found, scholars tend to view the art as fundamentally religious, frequently referring to the caves themselves as cathedrals, chapels, and shrines. The French prehistorian Jean Clottes, arguably the world's most prominent authority on cave art, told the *New Yorker*, "Everyone agrees that the paintings are, in some way, religious. I'm not a believer myself . . . [b]ut *Homo sapiens* is *Homo spiritualis*. The ability to make tools defines us less than the need to create belief systems that influence nature. And shamanism is the most prevalent belief system of hunter-gatherers." However, nothing in the art itself calls for a religious interpretation. It is for the most part simply paintings of animals, so a deeper meaning is certainly a speculation, though not necessarily

an incorrect one. Another preeminent French prehistorian, Norbert Aujo-
ulat, doesn't claim to know what the art means and says, "[I]n my own
experience—I've inventoried five hundred caves—the more you look, the
less you understand."

The term "cave art" may even be a misnomer. It may not have been seen
as art at all. In his book *The Primal Mind*, Jamake Highwater maintained
that while creating art in the modern sense is a "complex, idealized and con-
ceptualized act," for primal peoples, "the relationship between experience
and expression has remained so direct and spontaneous that they usually do
not possess a word for art." Although Highwater, who represented himself
as being of Cherokee descent, was later revealed to actually be one Jackie
Marks of Eastern European Jewish ancestry,* the idea of "no word for art"
resonated widely and continues to do so. In the article "No Word for Art
in the Tewa Language—Only Meaning," Greg Lonewolf explains that "in
non-Indian terms, I'm an artist. In the Tewa world they say of me, 'He's
a very skilled person. He knows many things.'" So the ancients may have
thought of their paintings not as beautiful but as useful.

Cave painting was not the only art born at this moment in history. Judging by archaeological finds, humans all over the populated world seemed
to acquire other creative abilities at the same time. In southern Germany,
a number of caves have given us several "oldest known" creations by *Homo
sapiens*, including the oldest representation of a human, the Venus of Hohle
Fels, a busty woman carved in ivory; the oldest piece of zoomorphic sculp-
ture, a man's body with a lion's head called the Löwenmensch figurine; and
the oldest musical instruments, a number of flutes made from vulture bones,
swan bones, and ivory. All of these finds date from about 40,000 YA.

Of the flutes, the one discovered in the Hohle Fels cave merits further
exploration. It looks decidedly modern. It is of a five-hole variety whose

---

* When *Star Trek: Voyager* cast Robert Beltran, the son of Mexican immigrants, as
the Native American character Chakotay, the powers that be decided they needed
to bring on a first-rate Native American consultant to keep the character as authen-
tic as they could. So they spared no expense and hired Jamake Highwater.

five notes make up what we call the pentatonic scale, widely used in ancient times as well as modern, including works by Frédéric Chopin and Taylor Swift, two people not usually connected by the word "and."

Why did we get the flute before many other essential technologies, such as the eyed needle and the bow and arrow? Why does music have such a profound effect on us today? Cognitive psychologist Steven Pinker agrees that it is a mystery. In his book *How the Mind Works*, he points out that "as far as biological cause and effect are concerned, music is useless." He concludes that, "Music appears to be a pure pleasure technology, a cocktail of recreational drugs that we ingest through the ear to stimulate a mass of pleasure circuits at once." Another theory posits that we learn the patterns in music and that the brain constantly tries to predict what will come next. When we are right, we are rewarded with a little hit of endorphins. If this is true, then it subtly points to this being the time when we learned to predict the future.

Concurrent with the emergence of music and art, there was also rapid technological innovation. New techniques were used to make tools, and the tools themselves became more specialized. Antler, ivory, and bone were increasingly used to make ever more-sophisticated tools and jewelry. We frequently find artifacts from this period constructed from multiple materials sourced from widely separated locations, executed with multiple technologies.

One question we have not yet posed is: Why would we create art and music at all? What is its survival value? Why did we evolve to love art and music and those who create it? Why do rock stars have groupies? Geoffrey F. Miller, an evolutionary psychologist and the author of *The Mating Mind*, suggests that the humanities as a whole can be explained not in terms of natural selection but in terms of sexual selection. He says, in effect, that your mastery of the electric guitar will in fact make people want to mate with you. Academics have tested this hypothesis a number of times. In one study, a man with a guitar asked women for their phone numbers, then tried later without the guitar. In another one, women were asked to rate the attractiveness of a guy with and without a guitar. In a third, online friend requests were sent to strangers, varying whether the profile photo had a guitar in it or not. Time after time, the guitar radically increased success rates.

Miller points out that luxury goods are defined by their high cost, and so people who consume them are signaling that they are better mating material than someone shopping at the dollar store. He likens it to a peacock's tail, which he explains "is not just an arbitrary outcome of sexual selection. It's there because it's costly, which means only those fit, healthy, strong peacocks can afford to carry around those tails."

He goes on to suggest that art, music, humor, storytelling, and all the rest are the same. There is no survival benefit for mastering these skills, even though becoming proficient at them requires a good deal of time and energy. Miller concludes, "Those who invest the most energy, the most maintenance time, the most genes, into growing the trait, will attract the most mates . . . [A]rt, music and creativity are in there by design, as fitness indicators."

Based on the archaeological evidence, anatomically modern humans have been around for more than 250,000 years. Yet there is no representative art that's 250,000 years old, nor 200,000, nor 150,000, nor 100,000, nor even 70,000. Then, all of a sudden, presto, there are Lubang Jeríji Saléh and Chauvet. The difference between that moment in time and the two million years that *Homo erectus* spent doing little more than sharpening a rock couldn't be starker. The fact that the technology all came along at once, that the art emerged fully formed, suggests something dramatic happened.

The Awakening

To recap: anatomically modern humans are about 250,000 years old. But mentally, well, that's a different story. Fully modern humans (FMH) seem to have come along relatively recently, perhaps just 50,000 YA, as evidenced by the abundance of figurative art after this date and the complete lack of it before then.

This sudden and dramatic mental maturation that resulted in FMH has been widely observed for decades. It goes by a number of diverse names. Jared Diamond calls it the Great Leap Forward. Yuval Noah Harari terms it the Cognitive Revolution. The technical term is behavioral modernity, but it is also called the Symbolic Thinking Revolution, the Creative Revolution, the Upper Paleolithic Revolution, the Human Revolution, or even Becoming Truly Human.

What shall we call it? I had hoped to build a meta term in the form of an acronym of all these names, but their twenty leading letters offer but one vowel, and a *u* at that, so I have decided to choose a more poetic term, the Awakening.

The Awakening is a great mystery, perhaps the greatest one of all, because it's the story of how we became us. There are two questions we would like to answer: first, when and where did it happen, and second, how did it happen? The "how" question requires the heaviest lifting: How did it happen in multiple places seemingly at once? How did it happen so abruptly? And, most challenging, how did it happen at all? We'll tackle the first question here and the second in a few chapters.

First: When and where did it happen? There is no consensus answer, but there are competing theories, each of which creates a new question. To start with, 50,000 YA is probably not the right number. I use it because it most closely aligns to our oldest examples of representative art. That doesn't mean older art doesn't exist; new "oldest" artifacts are regularly found. So it may be 60,000 YA, or 100,000 YA.

Given all that, let's examine five "when and where" combinations and judge their relative merits.

1. **Africa/100,000 YA**

   It's generally believed that *Homo sapiens* originated in Africa, per-haps near the present-day border of South Africa and Namibia. Africa has more genetic variation than any other place on the planet, which is what one would expect if our species began there, since the rest of the world is populated by isolated offshoots. We have good evidence that biologically modern humans migrated en masse out of Africa between 50,000 and 100,000 YA.[*] If the Awakening had happened before we left Africa, then the questions of why so many cultures produced similar art at similar times would vanish: we took this culture with us when we left Africa. However, there is a flaw in this theory: There is no evidence for FMH in Africa any earlier than anywhere else in the world. You would expect the oldest cave art to be there, not in Asia, and the oldest carved ivory figures to be there as well, not in Europe.

   In fact, the oldest undisputed representative art found in Africa is much younger—cave paintings from about 30,000 YA found in

---

[*] It is believed that there have been multiple migrations out of Africa over millions of years: some to Europe who evolved into Neanderthals, others throughout South Asia. My use of the phrase "when we left Africa" has an emphasis on "we"—the people that were already, or were to become, FMH.

the Apollo 11 Cave in Namibia. A number of explanations have been offered for why no earlier representative art has been found in Africa. They are all plausible, but certainly not compelling, since they are mostly speculation. One theory is that there is such art, and it simply hasn't been found yet. This could be true given the relative dearth of such fieldwork in Africa. Another suggests that African cave art was executed in cave shelters, that is, extremely shallow caves, and having been exposed to the elements, it has long since vanished.

But even if this is true, there is still a problem with this chronology: If the Awakening happened 100,000 YA, before we left Africa, where is all the art more than fifty thousand years old *anywhere* in the world? If we dispersed as FMH more than fifty thousand years before Chauvet, before Lubang Jeriji Saléh, where are all the paintings—in Africa or anywhere else—dating from then? Why would we have hundreds of caves with representative art after 50,000 YA and zero from caves before then?

2. **Africa/50,000 YA**
If we don't have the artifacts to support any dates for representative art before 50,000 YA, let's use that date and keep the event in Africa. But this won't work either, because we have good evidence of FMH scattered around Europe, Asia, and Australia by this time. This also doesn't explain why the oldest artifacts in Africa are younger than in those other places.

However, this theory could be salvaged if you say, "Biologically modern humans left Africa 100,000 YA and they weren't FMH. Then, a little more than 50,000 YA, in Africa, the Awakening happened and almost immediately those FMH started looking wistfully at the horizon, wondering what was to be found beyond it. Then, they headed to Europe, Asia, and Australia."

3. **Europe/50,000 YA**
Could the Awakening have happened in Europe, not in Africa? While that would explain the overwhelming abundance of artifacts

found there along with their artistic mastery, we have to reject this theory as well since we have similar art in multiple faraway places that predates European art by thousands of years.

I include this possibility because it used to be universally believed to be true, based on the fact that we found magnificent art all over Western Europe and none anywhere else. It was speculated that the introduction of Neanderthal DNA into our genome fortuitously modified us in a new and special way, that is, the Awakening. This view was certainly understandable given what we knew, and the number of painted caves in Europe is still dramatically more than anywhere else. But the dating of the Asian cave art is undisputed and predates anything in Europe by a few thousand years.

4. **Everywhere at once/50,000 YA**

What this theory has going for it is that it matches the archaeological evidence the best. Similar, fully mature art seems to have come on the scene around the populated world all at once. However, this theory isn't able to explain how this is even possible. How could peoples with no contact with each other, separated by thousands of miles and wide waterways, all Awaken at once?

Maybe the archaeological record is simply distorted. Consider this: 50,000 YA was the height of an ice age, and there was so much glaciation that the oceans were four hundred feet shallower than they are today, meaning there was much more land than there is now. But even four hundred feet doesn't free up enough land to connect Asia and Australia, and yet we have good evidence of humans there at this time. Conclusion? That we were seafaring by that time, skilled enough to cross the open ocean to Australia.* If we were sea people, living by the coast, then that might further explain the dearth of early cave art and artifacts in parts of the world; they're

---

* It is debatable whether Australia would have been visible to the ancient mariners, or whether they could have gotten there by hopping from one visible island to the next. The difference is material for both psychological and technical reasons. Regardless, the voyage would have been around fifty or sixty miles.

there, just under four hundred feet of water. Traveling by boat would also explain the rapid populating of the world at this time, the similar cultures in far-flung places, and more. The problem is that there is no direct evidence for this, as well as the question of how we managed to all suddenly lose the ability to build boats of this sort for forty thousand years after.

5. **Unknown, but gradually evolved and spread/100,000 YA**
Perhaps there wasn't a Big Bang, an instant when some dramatic fortuitous mutation produced a human with our full range of super-powers. Maybe we evolved "slowly" over the course of fifty thousand years or so. Perhaps 100,000 YA or even earlier, humans began to slowly awake, and that process, over tens of thousands of years, eventually resulted in Chauvet and Lubang Jeriji Saléh.

As reasonable as this might sound, it doesn't fit the evidence all that well. We don't have a gradual blossoming of creative ability but an explosion of it. We don't have stick-figure cave paintings; we have nothing—then out of nowhere, we have magnificence.

The net of all of this is that we simply don't know exactly when it happened, or where it happened at all. Luckily for our purposes, we don't need to know that in order to understand *how* it happened. That's where the next few chapters take us.

That Voice in Your Head

How did something so dramatic as the Awakening happen? To answer that, we must distill it down to its essential characteristic. While its archaeological indicators are technology and representative art, those are just the detritus of a much more fundamental change in human cognition. Clearly, the people who painted Chauvet and Lubang Jeriji Saléh *thought* differently from *Homo erectus*.

I think the heart of the Awakening is the formation of language. I don't think we had language before it; rather, we were animals stuck in an eternal present. Language is an obvious candidate for what makes our cognition different from that of animals. We will explore animal languages shortly, but everyone can agree that human linguistic abilities are in a class all their own.

As we explore this question of the origins of language, a certain humility is required because, of course, no one really *knows*. Until recently, the question was virtually taboo in science circles, because it was thought that nothing scientific could be said on the subject. Many believe this is still true. However, now linguists think otherwise and believe there is real evidence that at least allows us to make educated guesses. The puzzle pieces they try to assemble to make these guesses involve genomics, archaeological finds, evolutionary models, brain science, and differences between modern languages.

Why did we create language? What is its initial function? One common theory is that we invented language so we could swap social information, that is, gossip. Cited in favor of this view is that—depending on your source—between 40 and 80 percent of our conversations are about the doings of other people. I don't find this logic particularly compelling. It would be akin to saying that we invented smartphones so we could play *Pokémon Go.* That may be what we do with our smartphones, but that didn't have anything to do with their creation. Others suggest that we created language for other purposes, such as conveying information and coordinating action.

But these are simply uses of language, not about why it came into being. Why was it created in the first place? Although we think of language primarily as a tool for communication, that isn't its primary function. Its primary function is thought. Before we could communicate with anyone, we had to think in language. It is fundamentally an internal mental construct, the method by which we think and reason. It is that voice in your head. Using language to communicate with others is simply a byproduct of thinking with it. In the next chapter, we will explore how we came to vocalize it, but for now, think of language as being solely internal.

In his book *The Possible and the Actual,* French Nobel laureate François Jacob emphasizes this internal origin story of language, writing that "the role of language as a communication system between individuals would have come about secondarily . . . The quality of language that makes it unique does not seem to be so much its role in communicating directives for action as its role in symbolizing, in invoking cognitive images." He adds, "[T]he versatility of human language also makes it a unique tool for the development of imagination. It allows infinite combinations of symbols and, therefore, mental creation of possible worlds." The last two words of that quote—"possible worlds"—include "the future."

If you think about it, thinking is probably what you use language for the most. Most of us have a running commentary on the world going on in our mind that uses language. It goes all day. "Did I remember to shut the back door?" "Is this leftover chili still good to eat?" In fact, we have trouble turning it off, even for a minute. And when we lie in bed at night trying to sleep, there it is, as we replay the day's events or anticipate tomorrow's in our head. And even when we finally fall asleep, language is still there, powering our dreams.

The idea that language is the stuff of thought is brought home clearly when you consider people who are born deaf. If they are taught sign language, that becomes the language in which they think. The part of the brain that governs hearing is used for processing signs. Likewise, those born deaf who have schizophrenia don't "hear voices"; rather, they often have hallucinations of disembodied hands signing to them.

What would your mental life be like if you were blind and deaf and never learned language? No one has put this into words better than Helen Keller in 1908 when she wrote about her life before her teacher came:

> I lived in a world that was a no-world. I cannot hope to describe adequately that unconscious, yet conscious time of nothingness . . . Since I had no power of thought, I did not compare one mental state with another . . . When I learned the meaning of 'I' and 'me' and found that I was something, I began to think. Then consciousness first existed for me . . . It was the awakening of my soul that first rendered my senses their value, their cognizance of objects, names, qualities, and properties. Thought made me conscious of love, joy, and all the emotions. I was eager to know, then to understand, afterward to reflect on what I knew and understood.

Cogito, ergo sum indeed.

Keller says she had no power of thought before she had language. She could not think nor compare one mental state to another, thus there was no future or past. You can hardly read those words without trying to imagine what that must have felt like. This is what I suspect being an *erectus* felt like. Granted, unlike Keller, they could see and hear; but without language, thought would be formless.

The Origin of Language

How did this mental language come about? How did we come to think in language and then express it externally in words? The answer to that question is hotly debated. You wouldn't necessarily expect this to be the case. It seems more like a "Should you open presents on Christmas Eve or Christmas morning?" kind of question. But it sure isn't. Broadly, there are two views. The first is that language is a *skill* that we learn, like playing chess. It is a technology we invented, and it must be passed down from generation to generation to continue. The second is that it is *innate*, like your sense of smell. You are born with it.

At first glance, it might seem like the former makes more sense. Kids learn to speak with a few words, then put them together into sentences, then make common errors of grammar, and so forth. It looks like the way you learn to play chess: you learn how the pieces move, play poorly, then get better. Those who hold this view generally don't believe anything like the Awakening ever happened. They see the development of language as occurring gradually over millions of years, beginning with a few grunts, then becoming more complex. Further, they have a higher opinion of *erectus* than I present here, regarding them as language users who controlled fire and were even seafaring. Linguist Daniel Everett believes this, maintaining that *erectus* showed signs of culture, and, according to him, "You can't have a language unless you have a culture, and you can't have a culture unless you have a language." I disagree, but I don't want to give the impression the question is settled.

I believe the evidence is far more compelling that language is innate. This theory is most closely associated with Noam Chomsky, the father of modern linguistics. He and his adherents believe that we are born hard-wired with language. This Universal Grammar, commonly referred to as UG, underpins not just human thought but all human languages. To be clear, that voice in your head debating whether to eat a second slice of cake is not Chomsky's UG. That voice is English. UG lives below it. It is the stuff of thought. Chomsky explains: "Something is going on deeper which we can't introspect into, any more than you introspect into the mechanism of vision, and that's the language of thought. And it's probably universal. It's hard to imagine how it could be anything else. There's no evidence for acquiring it."

What does he mean that we can't introspect into it? While we think in words, we also think in concepts, almost like another language buried deep below our native spoken tongue, hardly perceptible and seldom contemplated. That's where UG lives. When you are out for a walk and you see a dog, the thought "I saw that same dog yesterday" probably occurs not as a series of words in your head, but rather in a different form, almost ethereal. You just know that, and the form of that knowing is a kind of mental sentence. But you instantly know whether your thought is "I see that dog all the time" or "I saw that dog for the first time yesterday" even without it being formed into words you're aware of. That's UG. And yet we also think in our native tongue as well. The words you are reading right now are becoming thoughts, and if I were to ask you if the words "no" and "dough" rhyme, you can probably answer instantly without needing to say them aloud.

The evidence that the language capacity is innate is pretty compelling:

First, virtually all children learn to talk, even those with severe mental disabilities. Likewise, most adults have much trouble learning a new language. As linguist John McWhorter of Columbia University says, "All mentally healthy children learn to speak the language that they are exposed to within the first few years of life. We are all familiar with how difficult it is to learn foreign languages as an adult or even as a teenager, yet children acquire those same languages flawlessly with no conscious effort."

This fact alone sure seems to suggest that language is something we are born with, that there is a developmental period in which it's easy to learn to vocalize language in our native tongue, and that biologically we grow out of it at a young age. Some say that to learn a foreign language with no accent at all requires you to speak it by the time you are a certain age—sometimes as young as four years old—or you can't do it at all. What other learned skills are like this? If you don't effortlessly learn to play chess by age four, you can still learn it well after that age.

In fact, children aren't really "taught" language in anything like the way they are taught chess. They seem to be hardwired to pick it up. Interestingly, children are also not "taught" how to tell stories; they seem to do it naturally, unlike other related activities such as reading and writing, which come only with effort.

Second, there are cases where children develop their own languages, suggesting that language capacity is innate, not acquired. Consider the classic thought experiment: imagine that you dropped four babies on a desert island with no one else there. Would they eventually develop a language to communicate with one another? Or would they have to start over, evolutionarily speaking, and it would take eons for it to develop? Those who believe in UG believe not only that they would develop a language, but that it would have a grammar similar to that of all other human languages.

In support of this view, a few real-world cases are offered. Chomsky relays one about three deaf children whose speaking parents were told they absolutely must not teach them sign language, that the kids needed to learn to read lips. The parents didn't even gesture by pointing at things. As he told *Forbes*:

> The children played together—I think they were cousins. And when this was discovered, around the age of 3 or 4, it was found that they had in fact developed a language. When it was investigated, it was found that it had the properties of normal language for children their age.

Another famous case involved deaf children in Nicaragua, where, in the 1970s, there was no attempt to educate the deaf and no knowledge of sign language. Deaf kids generally didn't know other deaf kids, as deafness is relatively uncommon and geographically disparate. They would communicate with their families with a few simple signs in which they would mimic certain activities, such as eating. That was all they had in the form of language. But in 1979, Nicaragua opened a school for the deaf and tried to teach the children Spanish solely by lip-reading. It didn't go well. However, administrators noticed that on the playground, the kids signed to each other. While the kids understood each other, the teachers had no idea what they were saying. Sign language expert Judy Shepard-Kegl was called in and worked on decoding the language, only to find that it was structured like other human languages, with many of the common conventions of grammar such as verb agreement.

Author Steven Pinker sees this as evidence for a hardwired language capacity. He says, "The Nicaraguan case is absolutely unique in history. We've been able to see how it is that children—not adults—generate language, and we have been able to record it happening in great scientific detail. And it's the first and only time that we've actually seen a language being created out of thin air."

Adding to this is that roughly half of all twins experience a phenomenon called cryptophasia, in which they make a language that they speak only to each other. This isn't as persuasive as the Nicaraguan example, because the twins have a language and make another one, but it at least suggests an innate ability.

From a brain perspective, it doesn't really make any difference what the language itself is; what matters is that there is a structured grammar. Thus sign, spoken, and written language are all languages. Children born deaf become proficient with sign at pretty much the same time other kids learn spoken language. In the US, those in the South sign slower than in the North, suggesting a Southern drawl. If those born deaf are taught just how to read lips, and not to sign, it doesn't really "take" in the same way—they don't "think" in moving lips. It is suspected that such a language is too abstract and unformed for the brain to really latch onto and think in.

Finally, it's worth pointing out that the family parakeet or dog that hears pretty much everything that is said to the new baby never learns language itself. Not even a bit. There is a particularly disturbing story from the 1930s that drives this point home in vivid detail. It goes without saying—but I will still say it—that the story I am about to relate would be unethical by any standard today.

In 1931, psychologist Winthrop Niles Kellogg and his wife, Luella, had a ten-month-old son, Donald. They decided to adopt Gua, a seven-month-old chimp, and raise the two of them in exactly the same way to test the ol' nature versus nurture question. The hope was that the chimp would learn to talk along with Donald. Obviously, she didn't, despite the Kelloggs' best efforts. In fact, the problem was that Donald began to act more like a chimp, biting people and such. They ended the experiment and got rid of Gua. Donald went on to become a doctor, then killed himself at age forty-two.

Spoken Language

Language emerged primarily as a mental construct. That's mainly what we use it for today. Then we learned to express it to others. But why did we bother to start vocalizing that stuff in our heads when a good argument could be made that we would all have been better off had we just kept our mouths shut?

The answer to this question is somewhat lengthy and, oddly enough, begins with honeybees. If you think about it, your average bee isn't that smart. Heck, the smartest bee in the world isn't that smart either. But the colony . . . the collective colony *is* smart. For instance, when looking for a new home, scouts who have found a good candidate are able to pitch their idea to the other scouts, who, if persuaded, will go and check it out themselves. Then there is a bottom-up, error-correcting voting system wherein various possibilities for the new home are weighed, and once a quorum of about thirty bees reaches 80% agreement, the decision is made and communicated to the rest of the bees, who fly en masse to the chosen domicile. This behavior is even more remarkable when you realize that none of these bees are over a month old and none have ever done this before or seen it done, so it isn't a skill they've learned.

Another impressive behavior: your human body maintains a temperature of around 98.6 degrees Fahrenheit regardless of the temperature around you.

Incidentally, no one is exactly sure how you manage to do this. Beehives also maintain a constant temperature—of about ninety-five degrees. Whether outside it is a hundred degrees or zero, the hive maintains its temperature. The bees do this by the ways they cluster together, fan their wings, or fetch water to bring into the hive. But no bee is thinking, "Man, it's hot in here. Hey, everybody! Let's start flapping our wings and cooling this place down."

The hive is what is known as a superorganism. Superorganisms can emerge when creatures with highly differentiated roles within a self-contained community interact with each other. Eventually, the creatures in the superorganism can't survive on their own, and their individual behavior only makes sense when viewed through the lens of what is good for the group, not the individual. While each member of the superorganism has a short life span, the superorganism as a whole can live indefinitely. A bee, for instance, lives a month, but the hive can last a century.

Superorganisms are emergent entities. Emergence is a phenomenon in which a group of things takes on characteristics that none of the individuals have. This can be expressed as "the whole is more than the sum of its parts." Emergence is a mysterious phenomenon—one we are a long way from understanding.

Ant mounds are another example of a superorganism. There is much specialization going on: Some members take care of the young, others forage for food, and still others defend the nest. Our modern day-care workers, farmers, and soldiers might see certain parallels between that and humans. By one census, about 3 percent of ants seemed to always be working, while 72 percent worked about half the time. The final quarter were never observed to be doing any work at all, just sittin' around. Perhaps this, too, parallels humans?

Superorganisms display amazing emergent capabilities. Consider Latin America's leafcutter ants. One group of these ants collects cuttings of a certain kind of leaf and brings them back to the nest. The ants don't eat the leaves; they eat a fungus that grows on the leaves. The fungus is so specialized now that it can grow only when tended by the leafcutters, and in turn, the ants require this fungus to live. Once in the nest, a different kind of leafcutter ant that is much smaller than the rest tends the fungus crop. These

Early human tribes were limited to about 150 people, or so we think based on today's hunter-gatherers. Groups larger than this would have split up, exactly the same way that beehives split in the spring. It is reasoned that 150 is about the maximum number of people with whom we can have a meaningful or stable relationship. We call this limit, whatever the actual number is, Dunbar's number.

One hundred fifty people in tight communication with each other could display abilities no single person could. For instance, they could agree on a complex plan to take down a woolly mammoth. Since everyone would act in accord, it would be like a single mind with a distributed body.

Thinking of humans in this context, all the chatter we constantly do with each other makes sense: Just as we use internal language to think, the superorganism began using spoken language to think as well. All the conversations happening in the group, that is, the superorganism, are the voices in its head. I know that sounds a little kooky, like something a crystal-wearing hippie would say, but think of it this way: Imagine a large, modern city as a superorganism. There needs to be enough food in the city to feed the people. Who decides how much food to bring into the city every day? No one. Grocery stores and restaurateurs place orders for the next week or month. People also need to get around town. Who decides how many cabs and rideshare cars will be operating that day? No one. When you think of the city this way, it looks like the beehive. There are enough groceries and cabs not because anyone is planning it all out, but because of the nature of the interactions between the people. So all the conversations in the city are the voices in its head. It is a restaurant owner talking to the employees about what they ran out of last week and what they need more of for next week; it is a cabdriver getting a call from a buddy saying there are fares to be had uptown.

Thus, we started talking to become a superorganism because the superorganism is much more robust than the individual people. Over time, natural selection began operating at the level of the superorganism, as well as the individual. Eventually, our superorganisms turned into our cities. Cities are born, grow, and sometimes die off, like so many honeybee hives. And, as mentioned earlier, eventually the individuals in the superorganism can't survive on their own. How many of us could thrive after being dropped into the wilderness? I know I'm no Bear Grylls. We will return to superorganisms in Act III when we ask if all of our electronic gizmos are part of our superorganism or whether they are forming their own.

Compared with mental language, spoken language is a bit of a kludge. Why would I say this?

First, we do it with body parts that were designed for other uses. Lungs are undeniably built to breathe with, tongues to help eat, and so on. Those are the true natural uses of just two of the organs involved in speech, but we've recently adapted them for another purpose: talking.

Second, consider the rate at which you speak. In English, in casual conversation, it is probably about 120 words per minute. A superfast speaker can hit four hundred words a minute. But according to research by psychologist Rodney Korba, inner speech runs at a whopping four thousand words a minute. Finally, virtually everyone has had thoughts *they cannot accurately vocalize*. Anytime someone says, "I'm not saying this right," or "I can't quite put this into words," their internal language is thinking thoughts that their spoken "kludgy" language cannot quite express. In a way, this is a bit like your sense of smell. We can smell odors that we don't have words to describe. The palette of aromas that the nose can sense is vastly larger than the handful of words we have at our disposal to describe them. So it is with internal language.

Yet, as kludgy as it is, think about what a marvel spoken language is, as is your ability to use it. You probably know fifty thousand words and recall most all of them in a fraction of a second, even words you haven't thought about in a while, such as "quiver" or "nuanced." But they are all there, like arrows in a quiver, ready to be used to express your most complex and nuanced thoughts.

Even with all those words, language is still somehow really easy for us to use. We are naturally adept at it. We're not even aware of a whole host of linguistic rules that we somehow unfailingly obey. In his book *The Elements of Eloquence*, Mark Forsyth points out:

Adjectives in English absolutely have to be in this order: opinion-size-age-shape-colour-origin-material-purpose Noun. So you can have a lovely little old rectangular green French silver whittling knife. But if you mess with that word order in the slightest, you'll sound like a maniac.

We follow other rules without explicitly knowing. There is a reason that clocks go tick-tock and not tock-tick. It's the same reason that you don't ever dally-dilly or refer to someone as raff-riff. Our languages seem to infuse our minds in ways we aren't even aware of, and it is not clear exactly why. Language's effect on us is so subtle that there is even a theory called nominative determinism that suggests that people gravitate toward jobs that match their name; that is, Bakers become bakers.

Language is almost entirely self-referential—completely circular. All words are basically defined with other words. And some do double duty, having multiple meanings that we hardly notice since we are so good at judging meaning from context. Some words, contranyms, mean both a thing and its opposite. "Cleave" can mean both to adhere to and to cut apart, "left" can mean departed or stayed, and "dust" can mean both to remove the fine particles or to add fine particles to.

Spoken language can be superior to thought language in only one way: how it is experienced. This is because it is multisensory. Spoken words can be full of emotion and delivered with passion. Hearing a sobbing plea or whispered words of affection moves us. Spoken words can be almost musical, and the silence between them can be full of meaning as well. A slight pause at different parts of the sentence changes the meaning entirely. Spoken words can be beautiful in a way that is beyond the grasp of the thought.

Whether spoken or thought, hands down, the most amazing aspect of language is the infinitude of combinations of words—and thus ideas—that it affords. Think about this: There is probably not a paragraph in this book that has ever been written or spoken before. Yet you read it effortlessly, even though each word slightly alters the overall idea that a given sentence is writing upon your mind.

While there are an enormous number of languages—several thousand at least—a research paper published in *Science Advances* titled "Different Languages, Similar Encoding Efficiency" concluded that the spoken forms of all seventeen languages studied by the researchers conveyed information at the same rate, calculated at just under forty bits per second. Some languages are less efficient on a per-syllable basis than others in their ability to transmit information, but they maintain the same forty bits-per-second rate because their speakers talk faster. This could mean that this is the optimal, or perhaps maximum, rate that the brain can create or process speech. We are able to *think* faster than this, which suggests that processing spoken language is a pretty CPU-intensive task for your brain.

We are almost ready to return to the question of how the Awakening happened. There is only one open question still hanging out there: Do animals have language? To make the case that inner language is what gave us the ability to think about the future and is the core of the Awakening, it must be demonstrated that language is unique to us. If animals, any animals, really have language in the way that we do, then language—mental or spoken—isn't our secret sauce, the thing that lets us see the future and be so preeminent on the planet.

Nonhuman Languages

A ll pet owners, I suspect, have wondered what goes on inside an animal's head. What do they know? Do they have thoughts in the same way that humans do, with an inner voice of some kind? Does a dog look at something and think, "Arf, arf, arf, bowwow, arf"?

Animal language is a contentious topic for two reasons. First, people differ as to the standard to which animal communication must rise to be considered a language. Second, they disagree in the interpretation of what animals that are purported to have language are actually doing. Because of this, there is no real agreement on whether animals truly have language.

To figure out if animals have true language, we will need to get really precise about what exactly constitutes language. What are its core elements? It would be helpful to start with a definition, but there isn't one. This shouldn't be too surprising, since we also don't have consensus definitions of concepts such as life, death, love, family, home, intelligence, or any of the other things that are so pervasive in our lives that we just naturally assume we know exactly what they are.

To be clear, we aren't talking about communication. That is far too broad a term to be useful for our purposes. Language is *used* for communication, but communication is simply conveying information. The dog that bites you when you try to take its bone away just communicated to you very clearly his disapproval of your actions. But we wouldn't say that is language. In fact, it is often asserted that all animals communicate, and whether it is literally true or not, almost all certainly do. Ants do it with pheromones; fireflies do it with light; and African demon mole rats bang their heads on the tops of their tunnels. Even plants communicate with each other, often through odors they release. Neurons, which are alive but are neither plants nor animals, communicate through electrical signals, while bacteria get the job done with chemicals. Many animals communicate by vibrating the surface of the earth, water, or other substrate, and surprisingly, the most common form of communication of all is bioluminescence. But if we call all of this language, then we are saying that the slime trail that slugs use to communicate with each other is the same sort of thing as *Hamlet*. And if we are claiming this, then the word "language" is so broad as to be meaningless.

Let's look at it from the other side. If we all agree that humans have language, what are its essential characteristics? There are four of them, and *all* human languages have all four. You might imagine that somewhere there's an uncontacted tribe that has a simplistic language, able to convey only simple thoughts like "Me hungry," but there is no such human language, at least not of the eight thousand or so that we know of. Isn't that interesting? All human languages are capable of expressing an infinitude of ideas with great nuance and clarity.

First, language involves *symbols*. American polymath Charles Sanders Peirce said there are three kinds of signs. First there are icons, which are representations of the thing itself. Next are indexes, which directly indicate something, the way a knock at the door indicates a visitor. Finally, there are symbols, which are completely unrelated to the item itself. The word "sled"—whether spoken or written—has no direct relationship to a sled. Symbols make up the vast majority of language. There are a few exceptions in which icons are used, such as with onomatopoeia as well as the many sign language symbols that represent the objects themselves. The little floppy disk image that you use to save documents was once an icon but is quickly becoming a pure symbol as floppy disks exit the public consciousness. The use of symbols is pretty common in animals. The vervet's warning cry for a snake does not sound like a snake, nor is it an index of a snake. It is a pure symbol.

Second, language is *multilevel*. The best way to understand this is to think of a writing system such as written English. Letters can be arranged in different ways to make words, but—here's the important part—words can be rearranged as well. This gives you lots of variation with a minimum number of symbols. In spoken language, the first level is that words are built out of phonemes. Phonemes are discrete, perceptible units of sound in a language. English has forty-four, while some languages have just over a dozen, and others nearly a hundred. They are all the different sounds we make when we talk, such as the three phonemes of "cat": /c/ /a/ /t/. And these forty-four phonemes can be recombined into an infinitude of words.* Phonemes can be combined and *then recombined in different orders* to make unlimited morphemes that can make unlimited words. The word "tack" uses the same phonemes as "cat" but in a different order: /t/ /a/ /c/. That's the first level. The second is that the different words can be strung together as sentences using rules known as syntax, and then the same words can be strung together in *a different order* to make sentences that mean something completely different. "Both Amy and Betty don't like Cindy" can be rearranged a dozen ways at least to say completely different things, such as, "Don't both Cindy and Betty like Amy?"

In a *Scientific American* article called "The Emergence of Intelligence," neurobiologist William H. Calvin points out that while chimps have about three dozen sounds that mean about three dozen things, English speakers have about the same number of sounds—phonemes—that are used to mean an infinite number of things. He concludes, "No one has yet explained how our ancestors got over the hump of replacing one sound/one meaning with

---

* Technically, words are composed not of phonemes but of morphemes, which are discrete units of *meaning*. "Bookkeeping" has three: "book," "keep," and "ing." "Ing" can't stand on its own, but it has meaning, as in the act of doing something.

a sequential combinatorial system of meaningless phonemes, but it is probably one of the most important advances that took place during ape-to-human evolution."

That brings us to the third essential characteristic of human language, *productivity*. Because language is multilevel, a few sounds get you nearly infinite numbers of possible utterances. That part is just a numbers game. Productivity says that with language you can make up sentences that no one has ever said before and other people will understand you. Not because they have memorized all possible sentences you might say and what they mean, but because the language is rich and complex enough that new ideas are understandable.

There was a running gag on *The Tonight Show* back in the Johnny Carson days to find the sentence most likely never to have been uttered. They finally settled on, "Isn't that the banjo player's Porsche?". Productivity means that you can say a sentence like that, which no one has ever said before, and others will understand it. Then, when new things come along, we make up new words or combine old words in new ways to create completely new things. Mail used to be in an envelope, then along came something new: email. Email wasn't a thing, then one day it was. Phones were phones until *cell* phones, then *smart*phones came out. In contrast, think of animals' warning calls to each other in nature. They make a certain sound, and it indicates a certain kind of predator is in the area. That works like a charm. But what if one of those animals just started shouting stuff none of the others had heard before? His buddies would be like, "What's up with Sid? Did he eat fermented berries again?" They wouldn't know what to make of it. Their sounds are not productive.

Finally, there is *displacement*. This is the ability to refer to things that are not currently seen, such as other places, times, or even imaginary objects. This allows us to expand our linguistic universe beyond the spot where we are standing and requires a certain amount of abstract thought. That's why we can say things to our friends like, "Do you remember that time when we were kids back on the farm and we tried to build a bird suit to fly in? We should try that again."

With those four features, let's look at some different communication systems and see whether they qualify as languages. It should be noted at the outset that many things that are referred to as languages are not. Math and music have syntax and symbols, and are often referred to as languages, but they aren't. They are simply notational systems and calling them languages is just a metaphor.

Let's start with plants. They do some pretty impressive things. Take acacia trees, whose leaves are a favorite of giraffes. When a hungry giraffe starts nibbling on an acacia's leaves, the leaves release a chemical called ethylene that wafts down to the other acacias, which, recognizing the alarm, begin emitting tannins that make their leaves toxic and foul tasting to giraffes. The giraffes, however, are on to the game: they start their leaf munching downwind and work their way into the wind toward the acacias that were unable to receive the alarm.

Are the acacias talking? No. We can anthropomorphize them by saying they are "shouting a warning" to the other trees about the giraffe scourge, but this isn't true. If the acacias have language, then they know only one word, and they can use it only one way.

That being said, some people believe that plants have language, a kind of sophisticated root talk going on that we are not privy to. Ecologist Suzanne Simard believes trees in a forest communicate with each other through fungi in the soil, where, she maintains, they "send warning signals about environmental change, search for kin, and transfer their nutrients to neighboring plants before they die." Perhaps this is true, but we are far from being able to say there is language. This could just be another simple communication process that we anthropomorphize using the metaphor of language.

But what of animals? Was Bertrand Russell right when he said: "A dog cannot relate his autobiography. However eloquently he may bark, he cannot tell you that his parents were honest though poor"? There are many claims: Bats are said to talk to each other, and they evidently bicker a lot. Parrots reportedly give their offspring names they keep their whole life. Elephants—which display all manner of complex personality traits, including mourning their dead—are believed to talk in a kind of sign language. Additionally, elephants recognize different human languages, avoiding those of people known to hurt animals. Of the many purported talking animals, it is prairie dogs, oddly enough, that I want to focus on. Let's look at the claims and the evidence.

Prairie dogs live in holes in the ground in large colonies. When they are out of their holes, they incessantly chatter with each other. When a predator comes along, one will sound an alarm, and they will scamper back into their holes. An animal behaviorist named Con Slobodchikoff wondered what was going on with all this chattering and scampering, and he has spent the past thirty years trying to sort it all out.

Early on, he noticed that prairie dogs had different calls for different predators—a hawk versus a coyote, for instance—presumably because they required different responses. Hawks are fast and need to be hidden from, while a distant coyote just needs to have an eye kept on him. Slobodchikoff also discovered that the prairie dogs varied the warning with information about the speed and direction of the threat.

Along the way, he recorded the animals' calls when they spotted him and his team, and they noticed that the calls varied a bit. He wondered why this was. He then had his students wear different color shirts and go out among the prairie dogs. He recorded the warning cries and believes he found that they basically said, "Blue human coming" and "Red human coming." Next, he got researchers with different body types and found that it was worked into the calls as well—"fat red human coming" or "short green human coming," and so forth.

Next, he rigged up a system where his researchers could pull a silhouette through the prairie dog colony. He used different shapes, like skunks and dogs, which elicited different warning calls. Then he mixed it up and used a shape the animals wouldn't know: a large oval, as if Humpty Dumpty were strolling through the colony. Sure enough, he maintains, he got a completely new warning call. He believes this is the attribute of language mentioned above: productivity. He also believes that the prairie dogs combine different signals in different ways (multilevel).

Linguists generally believe this is all crazy talk. They believe Slobodchi-koff is making two mistakes. One, he is assuming that what he is recording in the chatter are the essential elements of the prairie dog language—that is, that he has the phonemes right. While Slobodchikoff thinks such and such means "blue" in prairie dog, that's just him projecting. He is just seeing in the data what he wants to see: very human-looking adjective-noun-verb sort of language. But the larger criticism is that even if he is right, it isn't language. There is no underlying syntax, no way to go beyond the few simple combinations of alarm calls. Supporters of Slobodchikoff say that his critics just can't accept that his decades of meticulous field research have produced findings that run contrary to established linguistic thought.

Regardless, the prairie dogs don't seem able to talk about the past and the future, nor things that aren't there. Do any animals talk about things that are not in the here and now? The answer is almost always no, but there is one striking exception: the dance of the honeybee.

Some varieties of honeybee have a certain dance they do that tells the hive not only the direction of a distant food source, but how far away it is. That is essential information that lets the bees know how much flying fuel—honey—to ingest before making the trip. If that weren't enough, they also can communicate how good the source is, so the other bees can decide if it's worth the trip. These factors are conveyed using three independent variables: the angle of the dance relative to the (unseen) sun, the speed of the waggle, and the duration of the dance.

No one quite knows how all of this is accomplished, or even whether the bees are born knowing this or whether they learn it in their short four-week life. The former seems most likely, although the older bees tend to do the dance more accurately. And only a few honeybee varieties have this idiosyn-crasy, and the ones that don't seem to do just fine without it. Interestingly, bees on cocaine—yes, this has been studied—"talk" more and faster, and they usually exaggerate the amount of food at the source.

All this being said, this isn't a language. It has displacement in time and space, which is impressive. It also has symbols—the dance is not the nectar, nor does it look like it. But there is no productivity of any kind. They can't change the dance. Just like me in high school, the honeybees know only one dance move, and this is it. If they varied it even a little, it would be indecipherable to the rest of the hive. Humans have a hundred billion neurons in our brains. Honeybees have just a million. It is doubtful that the honeybee even knows what it is doing when it dances any more than a caterpillar knows why it is building a cocoon.

So far, no luck in finding animal language that shares even a majority of the essential characteristics of human language. But what about apes? We've all heard those stories of what they can do with sign language, but does this mean that apes have language?

Let's start with the claims. Certain great apes—chimps, gorillas, bonobos, and orangutans—are said to have learned vocabularies of human words, often by using variants of American Sign Language. Learning hundreds or even thousands of words, they are said to combine these words in new and novel ways, demonstrating the productivity characteristic of language described earlier.

One well-known example is Koko, a western lowland gorilla that is said to have learned a thousand different signs and could understand two thousand English words. She famously asked for a pet kitten, which she named All Ball. When the cat died and her caregiver Francine "Penny" Patterson broke the news to her, Koko signed, "Bad, sad, bad." Patterson also reported that Koko demonstrated displacement—talking about things that are not present—and could relay memories. She is also said to have made jokes and created words, such as calling a ring—a word she did not know—a finger bracelet.

Critics of Koko's language ability suggest that she didn't understand the signs *in and of themselves* but was conditioned to associate certain signs with certain outcomes. Supporting this is an observation about a signing bonobo named Kanzi. For Kanzi, signing is transactional—with 96 percent of his signs being functional, that is, to ask for things like food and toys. Only 4 percent are observational. Critics also point out that Koko and other signing great apes never demonstrate grammar (the basic rules of a language) nor syntax (the specific rules governing word usage and order).

Further, critics allege that most of the examples of great-ape language are interpretations of meaning that their caregivers have projected onto the words the apes are signing, the way pet owners will project human interpretations onto the actions of their pets, as in, "She thinks she's people." It is perhaps telling that often the caregiver is the only one who can translate for the ape because they know the animal so well. When Koko would get a sign wrong, critics claim that Patterson would say things like, "Stop joking around, Koko," assuming the gorilla knew the right answer but was just being uncooperative. In an *Atlantic* piece, journalist Roc Morin had a chance to have a fairly long visit with Koko. Toward the end of the piece, he wrote: "There was no way to know how much of her behavior was intentional and how much was my own or Patterson's projection. Allegations of selective interpretation have accompanied ape-language research from the beginning." Case in point: When another supposed loquacious chimp named Washoe saw a swan and signed "water bird," was she creating a word, as was claimed? Or simply signing the two things she was looking at, water and a bird? Morin went on to add, "Still, it was impossible to be there interacting with her, and not feel that I was in the presence of another self-conscious being."

But beyond all of this, critics claim there are a bevy of more philosophical reasons not to call what the apes were doing "language." Chomsky has said that if you can teach apes to use human language, you have effectively disproven the theory of evolution. Why? It is almost scientifically impossible that somehow language ability evolved in apes over hundreds of millions of years—and remains preserved to this day—even though the apes didn't know they had it. In other words, how could it have evolved, or been selected for, if it was never used for a survival advantage?

Chomsky explains: "If apes have this fantastic capacity, surely a major component of humans' extraordinary biological success (in the technical sense), then how come they haven't used it? It's as if humans can really fly, but won't know it until some trainer comes along to teach them."

Chomsky believes apes communicate with each other to be sure, but with regards to teaching them human sign language, he has this to say: "It's an insult to chimpanzee intelligence to consider this their means of communication. It's rather as if humans were taught to mimic some aspects of the waggle dance of bees and researchers were to say, 'Wow, we've taught humans to communicate.'"

In fact, in the examples of apes using language, Chomsky sees further proof that the language faculty is innate in humans. Human children learn language, grammar, syntax, and all the rest *seemingly without effort*, while the debate rages on whether any apes—even the very smartest—have ever formed a single syntactically correct sentence. With regards to the apes' use of symbols in general, the most generous thing we can say is that they can be taught to use them in a very rudimentary way, but no apes, or any animals for that matter, have ever invented them.

In a widely cited 2014 paper titled "The Mystery of Language Evolution," the eight authors, hailing from institutions including MIT, Harvard, and Cambridge, concluded: "Animal communication systems have thus far failed to demonstrate anything remotely like our systems of phonology, semantics, and syntax, and the capacity to process even artificially created stimuli is highly limited, often requiring Herculean training efforts."

Likewise, Yale linguist Stephen R. Anderson comes to a similar conclusion: "There is no evidence that any other animal is capable of acquiring and using a system with the core properties of human language: a discrete combinatorial system, based on recursive, hierarchically organized syntax and displaying (at least) two independent levels of systematic structure, one for the composition of meaningful units and one for their combination into full messages."

Do any animals, then, have true language? No. But before you accuse me of speciesism, let me point out that it is not clear that humanlike language and thought are good long-term survival traits. One proposed solution to the Fermi paradox—"Why aren't there aliens all over the place?"—is that intelligence leads to technology, which amplifies abilities, which the intelligent species uses to grow to unsustainable levels, displacing all other creatures and causing its own extinction quite quickly. From one vantage point, our intelligence and language look like an experiment of sorts, the results of which are not yet known.

How Did the Awakening Happen?

We are finally ready to address how the Awakening happened. It's a mystery for two reasons: The first is the disconnect between the archaeological evidence about our biology versus our abilities. Biologically, our brains had been growing for a long time—millions of years—but our abilities seen in the archaeological record remained flat until relatively recently, when they suddenly increased exponentially while our brains stopped growing. The second difficulty is that FMH are, for lack of a better word, miracles. Our abilities—complex language, knowledge of the past and future, creativity, technological mastery, consciousness, meta thought, and all the rest—so radically surpass those of any other creatures, including our own ancestors, that they defy explanation.

How did we come to have all these superpowers? Why do we have complex language, knowledge of the future and past, theory of mind, the ability to link causal chains, and all the rest? We have to squint just right to see vague shadows of these abilities in animals, who, even with great effort, cannot be taught how to do them, while we are born with them. Shouldn't there at least be some animals that have, say, 50 percent of our ability—or 10 percent? There doesn't seem to be an animal that can do these things even 1 percent as well as any human.

While writing this book, I mentioned these two problems to my wife, who said, "Let me guess. It was aliens, wasn't it?" The next day, I had a similar conversation with my literary agent, who replied, "Is this book going to be about aliens?" Both of them were jesting, of course, but the deep mystery of it all almost begs for an answer that dramatic.

The mystery of the Awakening reminds me of a passage in *The Lord of the Rings*, where Treebeard explains how trees came to talk: "Elves began it, of course, waking trees up and teaching them to speak and learning their tree-talk. They always wished to talk to everything, the old Elves did." This explanation would account for us as well. It is as if somehow the old Elves woke up a band of apes.

The Awakening seems so trippy that even tripping is offered as an explanation. A few decades ago, Terence McKenna, an ethnobotanist and professed psychedelic advocate, offered what became known as the Stoned Ape hypothesis: As our ancestors spread around the world, they invariably came to piles of animal dung with mushrooms growing out of it, and— it must have seemed like a good idea at the time—ate them. The mind-expanding experience made us what we are today. This theory has recently been revived by psilocybin mycologist Paul Stamets, who describes it as a "very, very plausible hypothesis for the sudden evolution of *Homo sapiens* from our primate relatives."

Although the Awakening is so big a question that it overlaps the fields of anthropology, archaeology, biology, psychology, sociology, neuroscience, religion, and philosophy, only three theories encompass all the non-alien possibilities: the Awakening happened one time to one person and spread from there; it happened to many people at roughly the same time through some kind of parallel evolution; or it happened gradually over evolutionary time.

One Person, One Time

Earlier, I posed the questions; "Where are the Bronze Age beavers? The Iron Age iguanas? Or the preindustrial prairie dogs?" If the Awakening was a freak genetic accident, a once-in-a-billion-years type of event, then that is the answer. We are the species that won the lottery. How might this have happened?

Evolution works via mutations in individuals, which, if beneficial, spread down to offspring and then throughout the species. Because of this, the thinking goes, the capacity for generative language, and all that comes with it, must have happened to someone once. For that person, the benefit was the language of thought. That person didn't have anyone to talk to, but they could *think* in language. Their thoughts had tenses, they could plan for the future, and all the rest. In a few generations, their entire tribe acquired this ability, which they used to communicate with each other. This band of humans became the primogenitors of us all.

This can be called the Chomsky position, although he isn't its only advocate. As he explains, from the empirical evidence we can surmise that "maybe seventy-five thousand years ago—some small neural rewiring took place, of course, in some individual, because it's the only possibility, and that individual had a computational process which was somehow linked to pre-existing conceptual structures." According to Chomsky, this produced "a language of thought. And then somewhere down the line it got externalized and you get interactions among individuals."

Chomsky believes that immediately after language emerged, over the course of a few tens of thousands of years, you see in the archaeological record "a sudden explosion of creative activity, complex social organizations, symbolic behavior of various kinds, recording of astronomical events, and so on. It's what Jared Diamond called a great leap forward. It's generally assumed, and plausibly assumed, by paleoanthropologists to be associated with the emergence of language."

Diamond does indeed agree. He points out that our DNA is virtually identical to that of a Neanderthal, but that we must have had some "Magic Twist" that made us so different. What was this Magic Twist? He writes, "Like some others who have pondered this question, I can think of only one plausible answer: the anatomical basis for spoken complex language." Or, as Yuval Noah Harari puts it, "The most commonly believed theory argues that accidental genetic mutations changed the inner wiring of the brains of

Sapiens, enabling them to think in unprecedented ways and to communicate using an altogether new type of language."

While each language has its own words and structures, all languages are at the core virtually identical. They all have, for instance, nouns and verbs. Linguist Joseph Greenberg identified forty-five characteristics that all languages share, or at least the thirty he used in his studies. This implies that language emerged once, fully formed, and it hasn't really evolved any since then. How can we infer this? Because there is, simply put, no such thing as a primitive language. Anywhere. Ever, as far as we can tell. And when we occasionally have first contact with one of the two hundred or so uncontacted peoples left on the planet, we discover that their language is as rich and complex as any.

So, if language emerged fully formed, then it emerged with object displacement, symbolic thinking, productivity, and with many levels. These elements of language both require and enable all forms of creativity, for our abstract language needs abstract minds, and vice versa. The one person this happened to was the first FMH. He or she passed this trait to future generations, and suddenly, in what would seem an instant, we were a creative, imaginative species, with minds and consciousness. We painted cave walls. We told stories. Our thinking transcended time, and our brains anticipated things. Being able to see the future and plan for it allowed us to become masters of the planet.

Critics cry foul at "all this happened at once" thinking, saying that evolution doesn't work that way. In a paper called "How Could Language Have Evolved?," four eminent academicians address this head-on. They reject the idea that evolution moves slowly and incrementally, saying that it does not fit the evidence "in which evolutionary change can be swift, operating within just a few generations, whether it be in relation to finches' beaks on the Galapagos, insect resistance to pesticides following WWII, or human development of lactose tolerance within dairy culture societies, to name a few cases out of many."

If this "one person, one time" theory is correct, it would mean that language evolved only once. Is there evidence for this? It used to be thought that several language families were created in total isolation and that our modern languages are each descended from one of these lines. For instance, Basque, spoken in parts of Spain and France, is thought to be unrelated to any other language. Some Basques believe that theirs is the language that Adam and Eve spoke in Eden. Other languages are similarly opaque about their origins and seem to be entirely original creations. But evidence now suggests that all languages spoken today do in fact come from a single mother tongue. This theory is largely based on the prominence of cognates in different languages. Cognates are words that have the same meaning and similar pronunciation in different languages. Often you can plow very roughly through a language you don't know because you recognize some of the words as similar to your own language. Those are cognates. In an essay called "Voices from the Past," eminent linguist Merritt Ruhlen wrote that "the significant number of such global cognates leads some linguists to conclude that all the world's languages ultimately belong to a single language family."

If this is so, it may be possible to figure out where this language was created. If, for instance, the word for "cat" is similar in many languages but the word for "dog" isn't, then it stands to reason that the mother language came from a place with cats but no dogs. By this reasoning, we can explain why the word for "mother" is similar in so many languages. It is because, well, everyone has a mother. By studying hundreds of cognates relating to weather, geographic features, plants, tools, and so forth, linguists have offered several candidate locations for where the mother tongue originated, many of which are in Eurasia.

Shigeru Miyagawa, a linguistics professor at MIT, sees it happening like this: "One way to think about this is that the brain, which had been growing ever larger for over a million years, at some point 75,000 to 100,000 years ago, hit a critical point, and all the resources that nature had provided came together in a Big Bang and language emerged pretty much as we know it today."

At first glance, this seems unlikely. Evolution never plans ahead; its time horizon is a single generation. An ever-bigger brain required ever more calories to support. Our brains are lavish consumers of energy, requiring about 25 percent of all our calorie intake, over twice that of chimps' brains. Human babies use nearly 90 percent of their calories to power their brains, so they are basically just a giant brain with a digestive tract. So growing our brains came at a high cost and would have lasted only if it offered some immediate survival advantage. Such an advantage couldn't have been associated with "smarts" as we think of it today, or we would see it in the archaeological record in the form of superior artifacts. But what if our bigger brain did have a survival benefit other than making us smarter? There are three things we do see in the record that line up timewise with our expanding brain. Our bigger brains could have:

- Provided us with the mental wherewithal to throw a spear at a moving target, an act that requires a large amount of computation that even today exists below our conscious mind. The anatomical changes that give us the unique ability to accurately throw overhand came along at the same time as our increases in brain size.
- Given us the advanced social skills we use to live in larger groups and work collaboratively, which we do much better than apes. Strong evidence suggests this tendency increased as our brains grew.
- Powered the complex mechanism that keeps our body temperature stable even after hours of work, another unique attribute of humans. No animal on the planet—not even a horse—can run a marathon faster than a human on a hot day, because we cool our bodies through a sophisticated mechanism of perspiration. This gave us a new way to hunt food: we just ran our prey down to exhaustion. If this last possibility turns out to be true, then a tip of the hat is owed to Aristotle, who always maintained that the brain existed to cool the blood.

For these changes to have endured, the growth of our brains must have accomplished something, but that extra wiring could also have been just one tiny mutation away from giving us our superpowers. Boston University's Andrey Vyshedskiy studies human imagination, and he and his colleagues discovered that children who have not been fully exposed to language when young—specifically before their prefrontal cortex matures—are never able to perform a kind of imaginative thought called prefrontal synthesis, which is the ability to imagine novel things. To account for the sudden emergence of creative thought in the world, he offers the Romulus and Remus hypothesis. It assumes a world where humans have speech, but they also have prefrontal cortexes that mature very early, before they can talk. Such a human would have speech but no imagination. If a mutation that simply delayed the development of the prefrontal cortex occurred in two kids, they would learn to talk before the cortex was fully developed and, presto, suddenly have a fully formed imagination, and stories flowed from that.

Could this ability have spread around the planet fast enough to give the appearance to archaeologists today that it happened all over the populated world at once? Probably so. A gene that allowed for adults to continue to digest lactose is thought to have become ubiquitous in certain areas in five thousand to ten thousand years, while a mutation that allowed humans to produce more red blood cells seemed to have taken only about three thousand years to become common at the high altitudes in what is now Tibet. Both of those spread quickly because they conferred a dramatic advantage to those who had them, while mutations that conferred only minor benefits seem to have traveled much slower, if at all. You can, for instance, still see old trade routes in the genetic bread crumbs dropped along the way for certain minimally advantageous traits that still have not caught on in the wider world. One can confidently assume that the mutation for speech would fall into "dramatic advantage" category.

If the Awakening really happened to just one person, then the roughly four billion adherents of the three religions that view Genesis as sacred text—Christianity, Islam, and Judaism—might see it being told in the story of Creation, when God "breathed into his nostrils the breath of life; and man became a living soul." That account begins with Adam and Eve, living as our ape ancestors largely did: unclothed, eating a fruitarian diet, and not practicing agriculture. Then they ate of the Tree of Knowledge and "the eyes of them both were opened, and they knew that they were naked." After that, they began to wear animal skins as clothing and went out and populated the world. Just as we did.

I believe the Awakening was likely a one-time deal with one person. But in fairness, two other theories deserve to be heard.

Many People at Roughly the Same Time

One day we are using a million-year-old stone tool, the next day we are painting on cave walls in Borneo, France, Germany, Central Asia, and Australia, among other places. We don't have any scientific explanations about how this could have happened. I don't even know any compelling pseudoscience ones. This is what the archaeological record *looks* like regarding what happened.

In *The Mother Tongue: English and How It Got That Way*, Bill Bryson remarks that we have no idea how groups of people who were scattered around the world "suddenly and spontaneously developed the capacity for language at roughly the same time. It was as if people carried around in their heads a genetic alarm clock that suddenly went off all around the world and led different groups in widely scattered places on every continent to create languages."

This sort of synchronicity doesn't just manifest in language, but in technology as well. Agriculture seems to have been invented in multiple places in a relatively short time by people unconnected with each other, such as in South America and South Asia. Writing seems to have been invented four times across two thousand years, completely independently. How do we know this? Because while all human languages are, at their core, quite similar, the writing systems invented at this time had nothing in common. For instance, Egyptians used an alphabet to spell words, while the Chinese used a massive number of symbols to represent complete words or ideas.

This continues into the modern age. A fascinating paper called "Are Inventions Inevitable?" from 1922 lists out 148 major inventions and discoveries that happened at the same time around the world. These include two men filing for a telephone patent just three hours apart, the discovery of sunspots in the same year by four men in four countries, and the invention of calculus.

How does this happen? There isn't really a simple answer. In general, similar environments produce similar results. Could something like that have caused the Awakening? If the Awakening was a biological event and not a behavioral one, then it's hard to see that happening without straying far afield of what we know of science. But, as Hamlet said, "There are more things in heaven and earth, Horatio, than are dreamt of in your philosophy." There is no end of speculative explanations, which of course could very well be true. To name a few, the Awakening could involve Paul Kammerer's seriality, Carl Jung's synchronicity, the collective consciousness, or James Lovelock's Gaia hypothesis.

Maybe the Awakening is somehow deeply coded into our DNA. Who knows what all is buried in there? I fully expect some day our so-called "junk DNA" is going to be revealed to be an encoded MP3 that begins with, "Never gonna give you up . . ." In all seriousness, consider the monarch butterfly and its four-thousand-mile, round-trip migration. No single monarch makes the trip—it takes five generations. But somehow, no matter where the butterfly is born, it plays its part in the journey. How is this possible? We aren't really sure. Monarchs seem to be able to determine both their latitude and their longitude, a trick we didn't master until the 1700s. Further, four of the five monarchs live about six weeks each, but the generation that makes it to Mexico, winters there, and heads back, lives six months. How is all this possible? We don't know that either. It might be the angle of the sun in the sky that activates genes that give that generation the unnaturally long life. Somewhere, somehow, it's all there in the base pairs.

Gradually over Evolutionary Time

A final school of thought is that there was no *ta-da* moment, that our Awakening was more akin to a night owl dragging themselves out of bed in the morning, staring at the wall while drinking hot coffee, and slowly waking up, only becoming fully human around 11 AM. Proponents of this view maintain that modern behavior slowly evolved in the earliest *Homo sapiens* in disparate pockets around Africa. They point out that in Africa we have found a few geometrically incised shells from 70,000 YA, and even older ochre—which may have been heat-treated to intensify the color and was likely used as body paint for either adornment or keeping mosquitos off—from perhaps as much as a hundred thousand years before that. They see a series of other incremental steps toward modernity that they believe date back to 250,000 YA. A new skill here, a new technique there, and so forth. The aggregation of all these abilities appeared to magically emerge in Europe and Asia 40,000 YA simply because that is exactly when we find *any Homo sapiens* in those places, as these FMH had just arrived there from Africa.

Conclusion

We are without a doubt all one people. In 1991, anthropologist Donald Brown wrote a book called *Human Universals*, which listed hundreds of "features of culture, society, language, behavior, and psyche for which there are no known exception." This list is huge. Just pulling from the *m*'s gets you magic, marriage, materialism, meal times, medicine, metaphor, music, and myths.

It seems unlikely that Brown's exhaustive list of universals would be quite so universal if they had evolved independently. So an image forms of a metaphoric Adam or Eve hitting the genetic jackpot, spreading their miraculous mutation to their tribe, who, over time, developed all those universals and then rapidly spread across the world, displacing all the other species in our genus: the Neanderthals, the mysterious Denisovans, the hobbit-like *Homo floresiensis*, and perhaps a few lingering *Homo erectus*. With their superpowers, they would have been unstoppable. And that's where we came from. That's us.

The Stories We Tell Ourselves

A t this point in our narrative, we have language but do not have stories yet. Stories are narrative constructs in which sequences of meaningfully related events unfold through time. Thus stories are built out of language, and language is required for stories. "But wait!" you may say. "What of silent films? Or what of the mime trying to escape the invisible box?" Those stories *do* require language; it is just internal to you. As you watch those, your mind provides the narration. During the silent film era in Japan, performers called *benshi* would narrate the action occurring on the screen, essentially giving voice to what everyone in the audience was thinking. Recall what Helen Keller said about her life before her teacher, before language, where there was no time, no discrete events, no cause and effect. Stories are not possible in that mental state.

Earlier we looked at how the primary purpose of language was not communication but thought. Stories are the same way. The real purpose of stories is mental. As Mark Turner writes in *The Literary Mind*, "Narrative imagining—story—is the fundamental instrument of thought. Rational capacities depend upon it. It is our chief means of looking into the future, of predicting, of planning, and of explaining. It is a literary capacity indispensable to human cognition generally."

The counterparts of these mental stories are what I will call "told stories," and they came along well after the mental ones. We'll talk about told stories in more detail later, but for now, when stories are mentioned, think first and primarily of the story occurring *in one's mind* as opposed to being told *around the campfire*. We'll gather around that campfire soon enough.

In the stories in our minds, generally we ourselves are the protagonists facing a problem we want to overcome. It can be simple, such as negotiating heavy traffic, or complex, such as planning an overseas vacation. If you should find yourself on a game show, deciding whether to take the $10,000 "sure thing" or risk it all for the $1 million prize, you very quickly tell yourself three stories. One where you take the sure thing, one where you go for the big prize and win, and a final one where you go for it and lose. You compare those stories in your mind and make a decision.

Stories must follow certain rules or they become incoherent. A protagonist, be it an individual or a group, has a goal. Events happen because of the actions of the protagonist or through actions of outside forces. Those events directly lead to the next events until the story's conclusion. Woven into the story are assumptions about what is good, preferred, desirable, or moral, and what isn't. So all stories—even the most mundane—are built on values of some sort. In addition, stories are about meaningful things changing. They have settings, context, and ultimately some resolution, which is the story's natural end. As Mark Twain put it, "A tale shall accomplish something and arrive somewhere."

Imagine you are driving in heavy traffic and there is an accident ahead. You want to change lanes to be able to exit and take the back way home. Your efforts to do this are a story, one that you might even recount to your spouse after you get home. And it has all of the elements mentioned in the last paragraph. Dressed up a little, it goes like this: "Once upon a time, a gallant prince astride his faithful steed found the path back to his castle blocked by a fire-breathing dragon. Thinking quickly, he recalled a nearby secret trail that the wizard had shown him as a boy and, taking the secret trail, arrived home to his fair princess, whom he regaled with the day's adventure."

We tell ourselves hundreds, even thousands, of these sorts of stories every day. Not the dressed-up version but simple, humdrum ones. Stories are the way we think. Let's say you are in the kitchen and hungry. You browse the contents of the pantry, considering multiple options very quickly. For each one, you tell yourself a really speedy story. First, you see a can of soup. You think about how you would prepare it. Stovetop or microwave? Then, you imagine how much you will or will not enjoy the soup. Are you in a soup mood? Then, remembering your waistline, you ask yourself how fattening soup is and whether you want to spend your "cheat" calories on it. Then you ask yourself if you have any crackers to add to the soup. You remember that you ate all the crackers last week, and crackerless soup does not appeal to you, so you move on to the next can. You just told yourself a mental story. Then there is a second one with the next can, the Wolf Brand Chili story. I know it doesn't sound like a big deal, because it is so effortless and the whole thing just takes a few seconds. Granted, Spielberg won't be optioning the rights to make a Tom Hanks film about your musings in the pantry, but nothing else on the planet can do this, and this seemingly inconsequential ability is the basis of our species' success. A trillion times a day, our species plans for the future. Compound that over time, and voilà! You have our world.

These simple mental stories enable us to live day to day. They fill our minds throughout the day, seamlessly connecting the past to the future, allowing us to live our lives constantly predicting the future, moment by moment. We do it across vastly different timescales, from considering the next few seconds all the way to centuries hence, as reflected in the adage that "A society grows great when old men plant trees in whose shade they know they shall never sit."

The ability to think this way was born in the Awakening and is *the* thing that makes us special. As Merlin Donald, professor of cognitive science at Case Western Reserve University, writes:

> The myth is the prototypical, fundamental, integrative mind tool. It tries to integrate a variety of events in a temporal and causal frame-work . . . Therefore, the possibility must be entertained that the primary

human adaptation was not language qua language but rather integrative, initially mythical, thought.

Donald is saying that the mind's primary job is to integrate, through time, sequences of related events, that is, stories. Our brains are fine-tuned for thinking in narratives. Social psychologist and author Jonathan Haidt said it succinctly: "The human mind is a story processor, not a logic processor." This would explain why, compared with computers, we humans are famously bad at all things quantitative and make all kinds of errors in logic, but we have an innate grasp of stories.

What Makes a Story?

For the kinds of complex mental stories we tell ourselves, language is the necessary but not sufficient requirement. There are three additional requirements to make a story. First, there is the notion of the past and the future, that is, an ability to think in tenses; second, an understanding of other minds—that each creature in the world has a different set of knowledge, beliefs, and goals; and finally, an ability to connect a series of meaningful related events into causal chains. Let's look at these.

The Future and the Past

Earlier, I mentioned that all languages exhibited displacement; they can refer to things not in the immediate vicinity. That includes other places and other points in time. For animal languages, that's a high hurdle, and few can reach it. But humans take the idea one step further in a way that's required for stories. What we're able to do is remember the past *as the past* and contemplate the future *as the future*. We understand exactly what the future and the past are. We know some events just happened, some happened before that, and others even longer ago. The same with the future. We perceive the timeline. Our capacity for mental time travel is so versatile that we can even think recursively about time. We can make statements like, "Yesterday

I predicted that by this time today I would be finished with the project that I have to turn in tomorrow."

Languages have difficulty dealing with abstractions such as time, so they compensate by adopting metaphors when describing them. For time, we use spatial references, such as a deadline being far away or our past catching up with us. Time can fly, drag on, or stand still. Usually, the future is in front of us and the past behind us. For the Aymara people of the Andes, this is reversed, with the future behind. Why? You have knowledge of the past, so it is in front where you can see it. The unknown future is behind us.

Imagine for a moment what it would be like *not* to be able to do this. Imagine if our thoughts were void of time. Not just thinking *in the present* but thinking without regard for time at all, like how a houseplant or a bacterium must exist. It is quite difficult—I sure can't do it—and yet that is how most living things on the planet think of time, while humans *constantly* think about the past and the future.

Stories—both mental and told—are always about the past or the future. They have to be. The present, the now, is tiny. It is the constantly moving 2.5- to 3-second window that neuroscientists say we perceive as the present. After three seconds, it becomes the past, and even one second from now is opaque. You can actually *feel* the now slip into the past: the next time someone is talking to you, notice how the words they said in the past couple of seconds have a realness, a depth to them, that the words they spoke even five seconds ago lack.

A stream-of-consciousness, present-tense narrative of your doings in that three-second window isn't a story, because the events themselves aren't meaningfully related. This is how a creature without time must experience the world. In his paper "The Ape That Captured Time," anthropologist Tok Thompson masterfully makes this point. He says that while humans may be the only storytellers, they are not the only narrators of the present. To support this, he offers the example of meerkats, which post sentries to be on the alert for danger. He writes that the sentries have "complex, socially-learned, and group-distinct calls for announcing approaching predators, as well as the 'all clear' signals once the danger has passed. The translation from any meerkat language would read something like 'Lion, distant. Lion coming closer. Run! . . . Now safe again.'"

The distinction between narrating the present and conceiving of the past and future is a crucial one. Thompson explains, "It is one thing to be able to warn of an approaching lion; it is quite another to be able to discuss lion attacks that occurred in the past, or that might occur in the future."

You might be tempted to argue that the meerkat told a story with a beginning, middle, and end. Many a *Dick and Jane* book has a simpler plot. But the meerkat didn't tell a story. You made that a story. You saw the events as being related to each other causally, and you inserted the timeline. The meerkat was just muttering what it saw in real time. Pretend for a moment that the meerkats had a couple of dozen different "words." Then, a meerkat might make this series of statements. "Hungry. Hungry. Loud noise. Danger? Strange smell. Sleepy. Hot. Find water. Ripe fruit." That time, the meerkat didn't get lucky and happen onto a story.

Consider how different our inner monologues are. A person faced with the challenge of getting their car out of a snowbank will tell themselves a set of stories about the future, each involving different strategies and scenarios to get out of the situation. "I can call AAA, but they will take forever to get here. Or I could call my brother-in-law who has a winch on his truck, but he'll call me a knucklehead the whole time he's here. I guess my best option is to pour gasoline on the car, set it ablaze, and walk away without looking back." Meanwhile, a meerkat in the same situation would think, "Car in snowbank. Push on car. Pull on car. Car in snowbank." The fact that our thoughts transcend time is one of the main reasons we dominate the planet. We imagine stories about the future, then actualize them.

Do animals think about the future and the past *at all*? One might assume so, since some animals stash away food for the winter or fly south when cold weather is approaching. But these behaviors are hardwired into the animals, not a function of their meteorological skills and prudent foresight. Animals do, however, exhibit other indications of time thought. A dog will bounce around excitedly when its owner pulls a leash out of a drawer, for the dog knows a walk is coming soon. But this is probably not a thought about the future per se but more like a single thing in the dog's mind, a leash-door-walk. Regarding dogs seeming to know when their owners are returning home or when dinnertime is, this likely isn't a sophisticated mastery of time but rather circadian oscillators, that is, the daily changes of neural activity and hormones that the dog experiences and associates with different positive reinforcement in a pure Skinnerian sense. These things just look like an understanding of time because that's how *we* think.

I remember, as a boy growing up on a farm, having to cross a certain field in which a particularly mean bull resided. The bull was faster than me, but I always took comfort that he could not plan for the future. When I would dash across the field, the bull would always charge at my *present* position, making constant course corrections as my position changed, resulting in his running in an arced path toward me instead of heading to where I was going. Despite being slower than the bull, I would always win that race.

Some experiments suggest that ravens can plan for the future. In one experiment, a certain human would give a raven a delectable morsel in exchange for a blue bottle cap, a pretty sweet deal for the raven any way you look at it. When offered a tray of enticing objects, the raven would take the bottle cap most of the time, even though the trading human was nowhere to be seen. Is this planning for the future? It looks like it because it is how we think, but some argue that it isn't, that the ravens simply developed a preference for the bottle cap. As corvid researcher Jennifer Vonk put it, "It isn't clear that this preferential selection reflects future planning."

But more and more studies seem to suggest that a few animals can plan for the *immediate* future. One showed that scrub jays would stash food in a foodless cage that they expected to be placed into the next day. Dolphins—nature's insufferable showoffs—seem to be able to plan for expected events a few hours away. And there is a particular compelling story of great apes' planning ability. The setup was this: There was an apparatus that contained visible food that could be opened only with a certain tool. When the apes were in the room with the apparatus, the tool was gone. Later, the tool was there but the apparatus was removed. Then, the apparatus was back, but the tool gone again. The apes quickly learned to take the tool when they left the room, so that the next time the apparatus was there, the coveted grape would be theirs. Even when multiple tools were strewn about, they would take the specific one that would later open the apparatus.

But in all cases, the skills seem to be limited and their time horizons always short. None of these animals would put money in an IRA. Compare the ability of these animals to that of the people charged with the construction of St. Peter's in Rome. In the interest of durability, it was decided that most of the art in the church would be mosaics, not paintings, and when ordering the glass tiles needed for them, the overseers required the artisans to deliver 10 percent extra of every color to be used in the centuries to come for inevitable repairs. Today, enough tiles are left to last another thousand years or so. Or how in 1830 the Swedish crown, still smarting from their naval losses in the Napoleonic Wars, planted three hundred thousand oak trees on the island Visingsö that would take 150 years to mature to provide timber for the fleet they thought they would need in 1980. They didn't need it, of course, and the trees are still there, but the Swedes planned for a future they expected beyond the lifetime of their great-grandchildren.

Meanwhile, in one study, when monkeys were offered the choice between one banana or two, they would take two. Who wouldn't? But when offered the choice between a few or many repeatedly, eventually they stopped caring. They were full; what did it matter? This suggests they weren't planning for a time when they would be hungry.

In his paper "Are Animals Stuck in Time?," psychology professor William A. Roberts asks whether the longtime assumption that animals have no sense of time is true and concludes "that the stuck-in-time hypothesis is largely supported by the current evidence." While acknowledging certain exceptions to this general rule, such as those cited above, he points out that if animals are stuck in time, then humans and animals are profoundly different. As he puts it, "Not only can humans travel backward and forward in time mentally from the present moment, but they can also contemplate what their cognitions about past and present were or will be at different times in the past or future. This temporal flexibility of cognition is vastly different from that of a creature that has no sense of time."

Memory of specific events is called episodic memory. The entire reason we have it must be to plan for the future, for it is hard to see any other use for it. What good is remembering the past if it doesn't help one survive and thrive in the future? Amnesia patients usually lose their ability to envision the future when they lose their memory of the past. Humans don't form episodic memory until around age four, and these memories are the first to go with diseases like Alzheimer's. But procedural memory—memory of how to do things—comes earlier and stays longer. The dog doesn't sit on command because it remembers being taught how to sit, but because it has become a procedural memory. Do animals have episodic memory, that is, a memory of a specific past and thus an ability to imagine a specific future? If so, it's minimal.

Beyond that, there is an additional limit to animals' episodic memory. Merlin Donald notes that while it is hard for humans to grasp, the truth is that "animals cannot gain voluntary access to their own memory banks . . . [They] cannot 'think' except in terms of reacting to the present or immediately past environments . . . [H]umans alone have self-initiated access to memory, or what may be called 'autocuing.'"

If you think about it, there is no reason that animals should be able to contemplate the past or the future. Neither of those two things actually exists. They are abstract ideas, not physical realities. That we can contemplate them is one of our many miraculous talents that we don't notice because it comes so easily.

Multiple Viewpoints

The next requirement of a story is knowledge of other minds—the understanding that each creature knows, believes, and wants different things than you do. This is a necessity for two reasons. First, for told stories, there really isn't a need to tell a story if everyone knows, believes, and wants all the exact same things. It would be like telling a joke everyone has heard a hundred times. Second, even in mental stories, acknowledgment that the various actors in the drama have different motivations and knowledge is essential.

In the earlier example of navigating traffic, the unspoken assumption in the mental story is that the other drivers have goals of their own as well as no knowledge of your goals.

This knowledge of other minds goes by several different names, including theory of mind (ToM), mental-state attribution, mind reading, perspective taking, and mentalizing. For the most part, I will use ToM for its brevity.

At what age do humans develop ToM? It's actually a pretty easy thing to test. You take a child and place her in a room, then have her dad come into the room and put a toy in a drawer. Moments later, have her mom come in, open the drawer, and place the toy somewhere else in the room, such as in a box. Then you ask the child, "Where does your dad think the toy is?" If she answers, "In the drawer," then it shows that the child understands that while she knows the object is in the box, her dad doesn't.

Until recently, consensus belief was that this ability developed in humans around age four or five and occurred in a certain order, beginning with knowledge that others have different goals, all the way to understanding that people may lie about their goals. Later it becomes awareness of sarcasm, hyperbole, and figurative speaking. But with the rise of the view of the innateness of language discussed earlier, scientists began to wonder if the knowledge of other minds wasn't also innate. Before, it had been assumed that children *learned* there were minds other than their own. But what if it's something we are born with?

Many experiments have been done along these lines, with ever-younger children, even before they have learned to talk. Using facial expressions and eye tracking as proxies for language, scientists have tried to re-create variants of this experiment. One study out of Hungary purports to show knowledge of other minds in seven-month-old babies. The setup went like this: Babies were shown one of four different movies involving a rolling ball, a rectangle, and a Smurf-like creature. One of the four had a situation where the Smurf thought the ball was behind the rectangle while the babies knew it wasn't, and that movie was the one the babies stared at the longest, suggesting at least some acknowledgment that the Smurf was surprised the ball wasn't there.

What of animals? Do they have a ToM? Not surprisingly, there is little consensus on this question because there are different interpretations of what certain animal actions mean. For instance, in her book *Spell of the Tiger*, Sy Montgomery tells the true story of a village being ravaged by man-eating tigers in the 1980s. Realizing the tigers always attacked people from behind, someone came up with the idea of people wearing human-looking costume masks on the backs of their heads. Nothing I've read specifies the particular masks they used, but for some reason I like to picture them as Richard Nixon masks. The plan worked perfectly: the tigers could never attack because either the farmer or President Nixon was always looking in their direction. Many villagers reported being stalked for hours by tigers, which even growled in frustration that the humans never turned around. For six months, none of the 2,500 people who wore the masks were attacked. But then, alas, the tigers figured out what was going on and started attacking again, but still from behind.

How do you interpret that story? Does the tiger not attack a person looking at it because it knows that the person can see it? Or have millions of years of natural selection hard-coded this attack-from-behind behavior into the tiger's DNA?

The question of whether animals have a ToM has been extensively studied, but the fact that the animals are nonverbal has necessitated a different approach to the question. One such setup tries to get at whether an animal can tell what another animal can and cannot see. If they realize the other animal can't see something they can clearly see, they must understand the other animal has different knowledge than they do. A few animals pass these tests, and apes do quite well on them.

However, there is another setup that stumps apes. In it, an ape sees a human placing food in one of several containers obscured by a screen. A second human then enters. Each human points to a different container. Does the ape then go over to the one pointed to by the person it saw hiding the food? It does not, suggesting it doesn't understand that one human knows something another one doesn't.

Or does it? It could well be the *pointing* that the apes don't get. Like virtually all other animals, they can't seem to tell what it means when we point to or look at other objects. That latter part—what we are looking at— is particularly interesting. Unlike apes, humans have substantial sclera, that is, the white portion of the eye. Why is that? In 2002, Hiromi Kobayashi and Shiro Kohshima offered the "cooperative eye hypothesis." It points out that with all that white in our eyes, we can easily tell what other humans are looking at. Our ability to do this is beyond amazing: from across a room, you can tell if someone is looking at your face or reading the witticism on your T-shirt. The hypothesis posits that as we cooperated more as a species, the white of our eyes became more important and was thus selected for. Apes are not particularly cooperative with each other, and thus never developed this attribute.

The mental abilities of dogs are a more interesting case. We have been selectively breeding dogs for their ability to know what we want them to do for forty thousand years. Dogs that could accurately interpret human desires became valuable assistants, while those that couldn't became dinner. That works out to over ten thousand generations of dimwitted canines being culled out of the gene pool, much more than enough time for natural selection to work its magic. We know this because the geneticist Dmitry Belyayev fully domesticated foxes in just forty generations, with substantial domestication occurring after just six.

Because of this long history, dogs have a unique relationship to people. One study found that a majority of Americans would end a romantic relationship if their dog didn't like their partner. This is in line with a finding of the American Academy of Matrimonial Lawyers that 96 percent of pet custody disputes are over dogs, while only 1 percent are over cats. Until quite recently, legally dogs were considered property and thus could be stolen, while cats were, well, just sort of there. And you know how yawns are contagious between people? Next time you are near a dog, yawn and often the dog will catch the yawn as well. This implies the presence of empathy, at least at a certain level.

A compelling case can be made that dogs have a ToM. First, unique in the animal kingdom, they effortlessly can tell what we are pointing at and will look in that direction. (Some studies suggest cats can do the same, but just choose not to. Yes, that sounds like a joke, but it isn't.) Additionally, a dog, placed in front of a food item that a human has forbidden it to eat, will obey, until the human turns their back. That suggests that the dog knows when the human can and cannot observe it.

Dogs pass another ToM test. Imagine this: There are two dog toys in a room. The human issues a command to the dog such as, "Bring it to me, boy." Which toy does the dog fetch? One study showed that if the dog could see that only one toy was visible to the human, it went for that one, whereas if the human could see both toys, the pick was random.

But detractors argue that none of this means the animals have knowledge of your mental states, rather that these are simply learned behaviors. Their interpretation of the examples above, which some studies seem to support, might run like this:

"Look, most often dogs can only perform these activities in their home environments with their owners, and the abilities they demonstrate are not really portable to other applications, suggesting they are not a generalized ability; rather, they are clever tricks the dog has picked up from living with you. In the 'two toys' example, if you take the dog into a forest and try it with two pinecones, the dog won't know what to do. Regarding the dog's waiting for you to turn your back to snatch the food, this is also learned. If you repeat the experiment with something else blocking your vision, say a red bandanna over your eyes, the dog won't get that you can't see it. Finally, a dog being able to discern what you are looking at and act accordingly doesn't mean it knows what you are thinking, rather that it has learned that eyes that do X should be followed by the dog doing Y. You can train a dog to do different things depending on your hand gestures, but that doesn't mean that the dog thinks your hand has a mind or a point of view."

One interesting aspect of animal behavior that implies a theory of mind is that animals play games with other animals of their species that often mimic hunting or fighting. They aren't really fighting, and they know the *other participant also knows they aren't really fighting.* Indeed, captive animals will engage in play with animals of other species they would normally never come in contact with, as well as with humans—a trait that presumably could not have been selected for over a long period.

What is perhaps most telling about animals' true abilities is a really astonishing fact: no animal has ever asked a question. Apes, which can learn hundreds or even thousands of signs, have never used them to ask a single question. None.* This fact could be interpreted to mean that the animal doesn't think we know anything that it doesn't already know. Questions to others imply a theory of mind.

This dearth of animal questions is striking because humans, especially children, ask so many. Paul L. Harris, a professor of education at Harvard, estimates that between the ages of two and five, a child who regularly interacts with a caregiver will ask forty thousand questions. Harris goes on to characterize those questions, stating that "about 70 percent of them are seeking information as opposed to things like, for example, asking permission. And then when you look at those questions, 20 to 25 percent of them go beyond asking for bare facts like 'Where are my socks?' Children ask for explanations, like 'Why is my brother crying?'" That is, questions whose answers are stories.

With an understanding of multiple viewpoints, humans are able to organize the goals of the various characters in their mental stories and use that as a framework for making plans, the way a socialite might assign place settings at a dinner party. "We just can't seat the bishop near the mayor because she hates organized religion. Seat him next to the duchess who recently lost her beloved Pekingese so he can offer her consolation." This seems so natural to us that it passes without notice, but this ability is unique to us, and in it might lie the roots of human consciousness. Consciousness is the experience of being you. A thermometer can measure temperature, but it cannot feel warm. That feeling, that experience, is a

---

* There may be one exception. Upon seeing himself in the mirror, a gray parrot named Alex asked what color he was, but that's it.

huge mystery. We don't know how it came about, what its purpose is, or even how mere matter can *experience* anything at all. It is the great scientific question we don't know how to pose scientifically—nor even what the answer could look like.

It might just be that our ability to recognize other minds created the inner dialogue that we have running in our heads. We think, "Well, I know this, but she knows that," then imagine what it would feel like to be her. This in turn might give rise to empathy, the abstract experiencing of the pain or joy of others. I don't offer this as a theory of consciousness—it is far too thin a broth for that—but perhaps our understanding that others have minds different from our own might have set us on the path to consciousness.

Causal Chains

That brings us to the third and final requirement for stories, both told and mental: an understanding of causal chains.

Consider this short story: "The dog barked loudly as the price of eggs in a distant city went up a nickel. Bob wondered if Susan knew that blue was his favorite color. Then the electrical grid in Tampa went down for a few minutes. Jill shook her head slowly as she decided not to participate in Taco Tuesday."

Okay, this is an awful story. No wonder my novels never get published. It is almost painful to read because your brain struggles to decipher how all these things are related. That's what human brains do. They stitch together events into causal chains; they make a story.

There is no limit to the number of causal events that can be strung together sequentially, nor to our capacity to follow those events. The heft of the Lord of the Rings trilogy and the Harry Potter series demonstrates that we have little trouble remembering long series of intertwined events. You probably don't reread "Jack and the Beanstalk" on a regular basis but could likely still recount the causal string: poor family needs money, sends Jack to sell the cow, Jack trades the cow for five magic beans, angry parents throw beans out the window, and so on.

The fact that our brains can do this ad infinitum makes us unique on this planet. You may be able to train your dog to fetch you a beer from the fridge—which involves several correctly sequenced steps—but our ability to sequence hundreds or thousands of things together is unequaled. Even speech itself demonstrates this: Every sentence you utter is an exercise in sequencing words together in a certain order to impart specific meaning. Our brains seem uncanny at doing this. But why? Why is this so important?

You can recite the alphabet, but to recite it backward, well, that's difficult. You probably know the eight notes in the do-re-mi series, but can you list them in reverse? Most of us have to say them forward in our minds a couple of times to be able to try to recite them backward, suggesting that they are stored in a certain order and can only be accessed in that order. This is no doubt for the best. If your recollections of the sequences of events were random and you had to mentally sort them out, that would get old really quickly.

These two facts—that most stories are told in temporal sequence and that the brain is hardwired to remember things sequentially—are intertwined, and when that is coupled with our knowledge that the future exists, it gives us our ultimate superpower: we can predict what will happen in the future.

We do it constantly, all day long. In fact, it is hard to *not* predict the future. If you are chatting with a slow talker, the urge to complete their sentences can often be hard to _____. See? You probably couldn't help filling in that blank. In his book *Stumbling on Happiness*, Daniel Gilbert doesn't exaggerate when he points out, "Most of us do not *struggle* to think about the future because mental simulations of the future arrive in our consciousness regularly and unbidden, occupying every corner of our mental lives." In *How to Create a Mind*, Ray Kurzweil doubles down on this, stating that "we are constantly predicting the future and hypothesizing what we will experience. This expectation influences what we actually perceive. Predicting the future is actually the primary reason that we have a brain."

The primary reason we have a brain. That's pretty big stuff, but I think it is spot-on. Philosopher Daniel Dennett brings the point home: "Learning depends on being able to extract information from your past and apply it in the future. All of life is a matter of exploiting the past to anticipate the present or the future."

We start predicting the future at a very young age. Perhaps we are even born doing it. A baby, just a few months old, will react with surprise if you let go of a ball and it just hangs in the air. They were expecting it to fall, and it didn't. Odd. Other studies show that when the expected behavior of water—such as being able to be poured through a screen—doesn't occur, the babies stare in apparent disbelief.

In his book *The Science of Storytelling*, Will Storr points out that we don't just try to predict the future, we also try to understand why things happen the way they do. He recounts one experiment in which wooden blocks were modified with weights to make them behave in strange ways. When kids, then later chimps, played with them and encountered their strange behavior, "most human three-to-five-year-olds inspected their bricks curiously in an attempt to discover the cause of the unexpected behaviour. Not a single chimp, in the same experiment, did. Humans, the professor of education Paul Harris has said, 'probe the how and why of things, sometimes tenaciously, even if it yields no tangible rewards.'"

This is a large part of how we came to develop not just science but imagination itself. Daniel Gilbert waxes poetic on this topic as he writes, "but to imagine—ah, to imagine is to experience the world as it isn't and has never been, but as it might be. The greatest achievement of the human brain is its ability to imagine objects and episodes that do not exist in the realm of the real."

On a day-to-day basis, we use these abilities primarily to make *plans* about the future. We draw upon our brain's vast library of episodic memories, which are saved as stories, to make plans. And plans themselves are hypothetical stories about the likely results of future actions. Thus we use stories of the past to tell ourselves stories about possible futures. What we do here seems to be quite different from what animals do. I've burned my hand while cooking any number of times. I don't remember any in particular, but I remember I want to avoid doing it. Other things, I remember all the specific instances, like the birth of each of my four children. The latter are episodes. All animals remember are the first example, the meta knowledge. They don't seem to be able to draw upon specific episodes when contemplating action. As polymath Roger Schank put it, "Figuring out how to behave in a new situation is most certainly helped by being reminded of an old situation that is like the new situation. The old situation then becomes a guide to follow or even a guide to what not to do. But all this depends upon finding something relevant to use as a guide in the first place."

There are limits to our ability to do all of this. In a sense, there is no such thing as a true story. Hollywood gets this, which is why films are "based on a true story" or, even further from the truth, are "inspired by real events." No story is truly true, because all stories are incomplete. They are editorial selections. At any given moment, you have a constant stream of data coming into your brain via your senses. You connect various bits of that data based on what you believe to be meaningful. Thus every telling of a story—even a mental story you tell yourself—is an interpretation of an abridgment of a narrative created in the past that can never be re-created.

The mental story is thus an ephemeral thing, and it feels nothing like reality itself. This is for the best; otherwise we wouldn't be able to tell what is real or not. We would be cursed to relive our worst moments again and again. Yes, our best ones are lost as well, but even they are, at best, "based on a true story."

Our ability to connect sequences of events together into stories and use those to make plans about the future is so robust that we see connections between events that aren't actually connected. This is known as the narrative fallacy.

Pigeons, as psychologist B. F. Skinner demonstrated in 1947, are superstitious birds. More superstitious than even baseball players. Skinner rigged an apparatus that would feed them—the pigeons, not the ballplayers—on a regular basis. Over time, three-fourths of the birds developed some kind of ritual—perhaps turning around three times quickly—that they believed would conjure up the beloved food-giving contraption. Of course it didn't, but whatever superstition they adopted must have happened once or twice just by coincidence, and the pigeons came to believe that that action caused the food to appear and would do so again. If pigeons, with their limited ability to connect sequences of events together, succumb to the narrative fallacy, what of us?

As Nassim Nicholas Taleb writes in *The Black Swan*, "The narrative fallacy addresses our limited ability to look at sequences of facts without weaving an explanation into them, or, equivalently, forcing a logical link, an *arrow of relationship*, upon them. Explanations bind facts together. They make them all the more easily remembered; they help them *make more sense*." That's all well and good, though if the narrative is incorrect, then not only are we wrong, we are simultaneously confident we are right.

Adages are useful things. They allow us to distill the experience of the ages into easy-to-remember nuggets that we can use to help shape the decisions we make. But here's the thing. They are often contradictory. Should you "look before you leap"? Perhaps, but what of "he who hesitates is lost"? Or how about "nothing ventured, nothing gained" versus "a bird in the hand is worth two in the bush"?

The problem is that we force those axioms onto the past to make sense of what happens. Books like *Good to Great* and *In Search of Excellence* do this in the extreme. They purport to explain why some companies succeed and others fail, and the explanations are always quite simple and straightforward. The problem is that business advice also exists in sets of dueling dyads of contradictory axioms. If a company changes with the times and goes out of business, we say they "didn't stick to the knitting," a famous phrase from *In Search of Excellence*. If they stick to the knitting and go out of business, then we say, "They didn't change with the times."

We want to believe that the world is a predictable place, and that success in life or business comes from following a few simple rules. But all these rules are really just explanations of why various past stories turned out as they did. The unsettling truth is that sometimes you should look before you leap, and other times, he who hesitates is lost. Wisdom comes in knowing when to do which.

We probably came by narrative bias for good reasons. Seeing patterns where there are none may just be the price that we pay for spotting patterns at all. Historically, the price of false positives vastly outweighs the price of false negatives. Imagine our ancestors getting a glimpse of movement through some thick brush. If they think, "That sure looks like a lion," and they are wrong, well, no harm done. But if they miss the pattern completely and it turns out that is a lion, well . . . you get the idea. The overly sensitive cognitive function that lets us see shapes in clouds or a face on the moon really is just another form of the narrative fallacy—seeing patterns where they don't really exist.

Biographers are notorious indulgers in the narrative fallacy. They almost have to be. They are signing up to summarize the seven hundred thousand hours of a person's life into a couple of hundred pages to explain why they turned out a certain way, and by golly, the reading public demands a straightforward, commonsense explanation. The only people more susceptible to the narrative fallacy are autobiographers—that is, ourselves—for, as George R. R. Martin puts it, "Nobody is a villain in their own story. We're all the heroes of our own stories."

By knowing these three things—that there is a past and a future, that things happen in causal sequences, and that other minds exist—we were able to think in a wholly new way, a way that gave us the edge to go from an endangered species (there weren't all that many of us) to masters of the planet.

Told Stories

Let's shift away from the idea of mental stories into what we more commonly think of as stories. We begin by returning one last time to animals and asking if they tell stories. Anthropologist Tok Thompson addresses this question sympathetically. He points out that until recently, at least in the West, the question would "have seemed ludicrous." He explains that in light of recent advances in understanding animal language and culture, the question becomes less crazy. However, he concludes that "the answer, after a careful review of data, appears to be no. While . . . nonhominid animals have not been demonstrated to tell stories, what is remarkable is how very close they come."

Animals' inability to tell stories shouldn't be much of a shocker for all the reasons we covered earlier: limited language, lack of theory of mind, lack of knowledge of the future, poor episodic memory, and all the rest. An additional deficiency would also be an inability to reason counterfactually, that is, speculating about how past events could have turned out differently.* What we conclude from this is that storytelling isn't some sort of basic

---

* An illustration of the kind of counterfactual thinking that humans do is the explanation of why Olympic bronze medalists are happier than silver medalists. The former thinks, "Whew! If I had gone just a little slower, I wouldn't have medaled at all," while the silver has a similar thought about how close they were

ability, but rather something complex and nuanced that requires a palette of special abilities to perform.

We, on the other hand, start telling stories soon after we learn to talk. Little kids are natural storytellers, who around age three begin spinning rich, imaginative tales of their own creation. In her book *The Stories Children Tell*, Susan Engel makes the following observation: "Between the ages of 1 and 8, children travel an extraordinary path, from uttering their first words to becoming complex and sometimes avid storytellers." But then, interestingly, much of this ability is often lost by age nine. Imaginary friends stop popping by, and the Dr. Seuss–like creatures that occupy their imagination vanish. Games of make-believe stop. This early acquisition suggests that storytelling comes from somewhere deep inside the child, perhaps in the base pairs of their DNA. They certainly don't—one hopes—learn about imaginary friends by watching their parents' interaction with their own, nor do they see their parents having make-believe tea parties with their friends. Of course, this natural penchant for fantastic stories doesn't completely vanish in everyone. Luckily there are enough J. K. Rowlings and J. R. R. Tolkiens around to fill our minds—young and old alike—with wonder.

While we stop telling fantastic stories, we don't slow down in telling more mundane ones.

Susan Engel, mentioned above, asked her college students to track how many stories they told in one day and discovered that "they had told anywhere from five to thirty-eight stories in the day, and felt that, if anything, they had missed some."

Earlier I referred to told stories being the kind we share when we are gathered around the campfire. I meant it figuratively, but stories may well have originated around a literal campfire. Obviously, we can't know that, but the idea is more than mere speculation. In a paper called "Embers of Society: Firelight Talk Among the Ju/'hoansi Bushmen," anthropologist

---

to the gold. It's a useful capability because it gives us the mental power to project different scenarios into the future, imaging what "Future You" would want you to do today.

Polly W. Wiessner of the University of Utah details her findings about the sorts of things these hunter-gatherers of Namibia and Botswana talk about around campfires. She compared her notes of 174 day and nighttime conversations among the Ju/'hoan (also known as !Kung) Bushmen and discovered that daytime conversation was largely about economic matters and gossip intended to regulate social relations. However, at night, when economic activity was precluded, the topics turned from the cares of the day to storytelling, singing, and dancing. She writes, "Night talk plays an important role in evoking higher orders of theory of mind via the imagination, conveying attributes of people in broad networks (virtual communities), and transmitting the 'big picture' of cultural institutions that generate regularity of behavior, cooperation, and trust at the regional level."

It's no wonder that people are different at night around fires. Think about it: Such gatherings often include people from both sexes and are multigenerational. The dazzling night sky stokes the imagination, and the impenetrable canopy of darkness that surrounds the fire just two dozen feet away in every direction brings a sense of community to the group. The fire itself is otherworldly; it is almost alive. It glows, it dances, it pops and crackles, and its warmth seems like the warmth of a living thing. Who would deign to discuss the price of corn in such a place?

What Wiessner found was that during the day, 6 percent of talk was about stories and myth. By night, that share rose to 85 percent. In her analysis of other forager cultures, she noted other examples of this broad pattern, such as the Ainu hunter-gatherers of Japan, who devote their days to their activities and the nights to deities and demons, as well as other cultures where "night was prime time for entrance into imaginary worlds of the supernatural."

The allure of the campfire seems to be embedded deep in our consciousness. Christopher Dana Lynn, an anthropologist from the University of Alabama, reflected on the effect that fire has on us and wondered if it was medically measurable. So he constructed a test in which he took the blood pressure of a group of people, stuck them in a darkened room wearing noise-canceling headphones, and played a video campfire on a twenty-inch computer monitor. Sure enough, down went their blood pressure compared with the control. Lynn tried to tease out what part of the fire had the mojo, and it turns out that the video, when played on mute, got nothing, but when fed the sound of the fire only (and no video), blood pressure went down, but not as much as video plus sound.

Just as the experience of hearing words can be superior to thinking them because of the multisensory aspect, a told story can be masterfully better than a thought one. Anyone who has ever sat at the feet of a great storyteller has experienced this firsthand: the setting of scene; the voices; the changes in tone; and often, surprisingly, the moments of silence. Live theater tries to capture this and has been trying for millennia. And today's films? With their hyperrealistic effects and mesmerizing musical scores, projected in brilliant color in a darkened theater? No matter how long I live, I will never think a story as well told as, say, The Lord of the Rings movies.

We've had writing for only five thousand years but speech for more than ten times that. This means that stories have been entirely oral for the vast majority of history. Everything humanity had learned over the eons had to be passed along verbally from generation to generation. All culture, all knowledge, all traditions, had to be handed down, usually in story form. A single break in that chain would reset all human progress back to the beginning.

One wonders to what extent stories change in their retellings when they are handed down over the ages. Are they reimagined each generation, like the continual reinterpretations of characters like Dracula or Sherlock Holmes? And what of the epic poems? Were they passed down word for word, carefully memorized to preserve their meter and rhyme? Or were they organic things, always evolving? Through the telephone game, we all learned how a single message can get distorted in retelling quite quickly. Is that how we should think of the oral tradition? Is it the telephone game played out over centuries? Or something more like sacred texts passed down with care and fidelity?

We don't know for certain because oral stories leave no fossils. But the oral tradition still lives today in every culture. If today's traditions are any guide, it is more like the Dracula example. The pieces are all there—the bloodsucking, the mystic powers—and the storyteller mixes and matches to create what copyright lawyers call derivative works.

Native American traditions are particularly interesting because, before European contact, they had no writing systems. They have relied on oral transmission for tens of thousands of years, right up to just a few centuries ago, making their cultures a good place to look for insights on that question. In A. L. Kroeber's 1948 book, *Seven Mohave Myths*, he recounts the kinds of variations he observed in listening to the Mohave storytellers, saying, "If the narratives are long, they almost inevitably show minor inconsistencies. The narrator may say that a thing is done four times, and then proceed to narrate six variations of it. Contradictions of plot may occur through lapses of memory or shifts of the narrator's interest." In a different work, Kroeber expands on this, likening Native people's storytelling to listening to a classical concert: You know the melody of the piece, but the interpretation of the performer is what gives it new life. He says that the tellers of the stories changed them up in the telling, "which endowed the old story with its special contemporary relevance. Alterations usually made the story particularly applicable to current circumstances, community issues, familial difficulties, new ideas about traditional practices."

Other ethnographers report similar findings regarding various cultures' core stories, the ones that recount and preserve their culture. The stories are often lengthened or shortened to suit the situation or needs of the audience. They are often episodic, and the different stories in a protagonist's life are told in various orders, or some are simply left out.

So is that the conclusion: that a culture's core myths are ever changing, reinvented with each generation? Life on Earth is billions of years old, but all living things are themselves young. Are stories the same? That, at best, only the broadest outlines are preserved, since the fallibility of human memory constantly morphs all details?

Maybe not. The people of antiquity undoubtedly had better memories than we do. In a world where you couldn't write anything down, if you wanted to know something, you had to remember it. Snippets of history offer tantalizing hints of people with memories better than our own, such as a Roman general who remembered the names of his thousands of troops, along with the names of their families. Or the story of Simonides from ancient Greece, who is said to have been at a banquet and happened to step outside for a minute when an earthquake hit. The banquet hall fell in, mangling all the attendees' bodies beyond recognition. Simonides closed his eyes and recalled where everyone had been sitting in the great hall.

Both stories are likely apocryphal, but there is no doubt that the advent of writing was believed at the time to be detrimental to memory. In *Phaedrus*, Plato has a character speaking of this very matter, stating that with regards to writing, "If men learn this, it will implant forgetfulness in their souls. They will cease to exercise memory because they rely on that which is written, calling things to remembrance no longer from within themselves, but by means of external marks."

However, perhaps even with their better memories, their lack of writing meant they could remember *nothing* verbatim. The late professor of psychology Ian M. L. Hunter believed that verbatim memory without written support is limited to fifty words in humans. He writes that "the human accomplishment of lengthy verbatim recall arises as an adaptation to written text and does not arise in cultural settings where text is unknown. The assumption that nonliterate cultures encourage lengthy verbatim recall is the mistaken projection by literates of text-dependent frames of reference." In other words, sure, you can memorize the Gettysburg Address but only if you have a text copy of it. Without that, the very idea of a verbatim recall is hard to grasp. If there is no definitive source, then what is the point of the exact retelling?

Even in our day and age, we may not be able to transmit oral accounts with fidelity, no matter how hard we try. In fact, your most vivid memories, so-called flashbulb moments such as where you were when you heard about the 9/11 attacks, may not be true at all. While this seems almost impossible to fathom, it has been demonstrated experimentally time and time again through the past few decades about key events, including Pearl Harbor, the John Kennedy assassination, and the *Challenger* explosion. Cognitive psychologist Jennifer Talarico was a graduate student at Duke when 9/11 happened. She and her adviser, David Rubin, saw this as a chance to study flashbulb moments, so the next day they got fifty-four students to record where they were and what they were doing when they heard the news. They also asked the students to recount a much more mundane recent memory as well. Following up later, they found that major discrepancies had crept into the students' memories of both events, at roughly the same rate. But the students' confidence that they were remembering the flashbulb moment correctly was incredibly strong, even when they were wrong. One study found that three years after the 1986 *Challenger* explosion, not one of the forty-two people being studied remembered the event the same way as they had earlier.

Oral tales do undergo a certain kind of Darwinian natural selection. The number of tales that *can* survive is finite; the number of potential tales is infinite. So if a story wants to live, it had better pass its memorability genes along to its offspring or it will go extinct. Rhyming is one strategy employed by poetry and ballads to be more memorable. Interestingly, epic poetry seldom uses rhyming schemes, perhaps because it could detract from the gravitas of the piece itself. However, it virtually always employs meter. The *Iliad*, the *Odyssey*,* and the *Aeneid* all use dactylic hexameter, the main meter of Greek and Roman poetry. Other tricks that stories in the oral tradition use to survive are remarkably effective. They eschew abstractions in favor of actions, usually involving powerful imagery: "Oh, Grandma! What big teeth you have!" and "The wolf huffed and puffed and blew the house down." They employ repetition, such as Goldilocks trying the various porridges, chairs, and beds, and they are frequently chanted or sung. Finally, the oral tradition may not have been purely oral. The stories, the important ones, may have been acted out, not simply told. Could this have been the purpose of those magnificent paintings at Chauvet and elsewhere? To be the backdrop on which stories were told?

---

* The *Iliad* and the *Odyssey* may be special cases. A group known as the Homerìdai from the Aegean island of Chios claimed to be at least symbolic "children of Homer" and maintained that Homer hailed from their land. They may have carefully preserved the writings of Homer, treating them with almost scriptural care. Or perhaps they simply continued the tradition of Homer or invented him altogether. There are entire books on this topic.

Why Tell Stories?

To recap: we use stories as mental constructs to think and to frame possible futures. That must have been their original purpose as well the dominant thing we do with them today. While we sometimes explicitly and carefully consider all possible future stories—this is called scenario planning—we do it implicitly throughout the day, from moment to moment. It is how we think. Contrary to what the mystics and New Agers say about the virtues of "living in the moment," that is not our way. We constantly reside in the immediate future, informed by the past and viewed as a series of connected causal events. In fact, because of how our brains are wired, this is how we must live. Kevin Kelly, eminent futurist and one of the cofounders of *Wired* magazine, told *This American Life* a story about how he tried for a time to live exclusively in the present and concluded that it was an "entirely unnatural and inhumane way to live. Having a future is part of what being human is about, and when you take away the future for humans, you take away a lot of their humanness. One needs to have a past and one needs to have a future to be fully human."

But here's the question: Why do we *tell* stories? It must confer some benefit, because the behavior has emerged and remained. It permeates our psyche so deeply that most children start telling stories within a year of learning how to talk. This is a bit odd if you think about it. Many stories are fictional, so they are, in essence, lies. And they are lies told to people who know they are lies. You would think we would have developed an aversion to them.

The answer is the same as why we externalize speech: it serves the super-organism that is our society. As we explored earlier, we as individuals think in stories by recalling episodic memories, that is, stories of our past, and then we use those to predict and plan for the future. Likewise, the stories that a group of humans, say a tribe of 150, tell each other are the way that the superorganism draws on episodic memories and uses them to predict and plan for the future.

Before exploring this further, let's give our superorganism a proper name. What shall we call an individual tribe of humans? How about Agora? In ancient Greece, the agora was the place in the town where everyone came together and transacted business, tended to politics, exchanged information, and swapped stories. Located either in the middle of the town or by the port, the agora was where it was all happening—a huge mass of people interacting with each other.

How does Agora think in stories? The same way people do. When we are contemplating engaging in some bad behavior, we consider it in light of a series of stories we have heard about similar situations. Agora does this but at the level of the group: as members of the younger generation grow up, they, too, are tempted by the allure of bad behavior. So tales with morals are told around the fire to keep such behavior in check. Stories of past heroic exploits are told to remind Agora of the reasons to be heroic in the future.

Thus told stories are, for the most part, Agora remembering the past in service of the future of the whole. This is a testable hypothesis. If it is true, then over thousands of years, we should see two things happen: first, as Agora grows and changes, the stories it thinks in its head will change as well. Our ever-changing body of stories should in some way map to world history. But second, counterpoint to that, we should see certain archetypal stories that never change, for they relate to the unchanging characteristics of all humans.

Let's spend a chapter on each of those.

The Timeline of Tales

Superorganisms like Agora have memories that in turn give them phobias and fears along with wants and desires. Can we see Agora growing and changing over the eons? I think so, at least in broad outline.

Let's break up the past into five eras and look at the stories we told in each.

Era 1: Newly Awakened (c. 50,000 YA to 12,000 YA)

Anyone who has ever camped far from cities knows that when night falls, a sort of magic happens. Everywhere around you, you see impenetrable inky black. But glance up and you will behold two thousand stars lighting up the night sky. The great band of the Milky Way, the cross-section of our galaxy, is painted across that canopy from one horizon to the other. It is seldom seen by most of us today, and so it is no surprise that in 1994 when a massive blackout hit Los Angeles, people called 911 to report that strange thing in the sky. But fifty thousand years ago, before light from our cities turned the night sky a deep and lifeless gray, our ancestors must have lain on their backs staring up at that vast celestial panorama and wondered.

What are the oldest told stories that survive to this day? That might seem like the sort of question we can't really answer, but fortunately, folklorists are an exceptionally clever bunch and can infer quite a bit.

For instance, consider a story from the Ket language, spoken today by just a very few people in Mongolia. A century ago, an Italian linguist named Alfredo Trombetti speculated that it might be of the same language family as Navajo. While this has not been definitively proven, subsequent research strongly suggests this is the case. There is even good DNA evidence that these two peoples are related. Think about that! Fifteen thousand years ago, give or take, some people from Asia crossed a land bridge into the Americas and populated a continent. And we know some of the words that they spoke, the ones shared by both Ket and Navajo. But that's not all they share. In traditional stories of Europe and Asia, the Big Dipper is associated with a bear. This is odd, because let's face it, it doesn't look at all like a bear. Furthermore, in many places, the ladle of the dipper is the bear's body and the three stars that make up the handle are the bear's tail, which is also odd because bears don't have long tails. In eastern Russia, however, the local folks know firsthand what bears look like and don't identify those three stars as a tail, but rather as three hunters chasing a bear. Additionally, there is a faint small star called Alcor near the middle hunter, which the Siberians see as a little bird helping the hunters track the bear. What's fascinating is that in many Native American cultures, the Big Dipper is also seen as a bear, being chased by three hunters with a bird showing them the way. It's the same story, and since these cultures became geographically isolated from each other, it must be a story that predates the crossing of the land bridge across the Bering Strait, some fifteen thousand years ago. We can be confident in this because if the bear interpretation came to North America after 1492 from Western Europe, then the handle would have been the tail, as was the European tradition. As Bradley Schaefer, a Louisiana State University professor emeritus and self-described "cheerleader" of this narrative, puts it, "We are very sure of the basic picture, even though we do not have signed affidavits all along the way."

Is that the oldest story? Probably not. It's just one we have some interesting data on. A more speculative theory offered by Australian academics Ray P. Norris and Barnaby R. M. Norris suggests that the story of the Pleiades star cluster as seven sisters being chased by Orion goes back a hundred thousand years before we left Africa. They point out that this story occurs in similar forms across the world, including among the Indigenous peoples of their own continent, who came into contact with Europeans only within the past couple of centuries. They argue, therefore, that the story was brought there by the original settlers more than fifty thousand years ago. Icing on the cake with this theory is that it offers a plausible answer to why we call them the seven sisters when for most people, only six are visible to the naked eye. It could be that a hundred thousand years ago, the two stars that appear to us to be a single one were perhaps just a little farther apart and thus discernible.

So the story fragments that come to us from Era 1 are from the night sky. Agora, wowed by the world it awakened in, tried to make sense of the place.

Era 2: City dwellers (12,000 YA to 4,000 YA)

Let's go forward a few thousand years to fairy tales, a surprisingly ancient form of story. They are defined as short stories that feature magical elements such as talking animals, witches, or, well, fairies. Ironically, they often didn't have "fairy-tale endings." Fairy tales are different from legends, which are generally believed to be true, or at least "based on a true story." Fairy tales are made up of archetypes, not characters, that are instantly recognizable to us, from the gentle and virtuous young maiden to the wicked stepmother.* The characters never have any nuance; they are good or evil, beautiful or ugly, and they have no depth at all. As author Philip Pullman observes in his *Fairy Tales from the Brothers Grimm*, "One might almost say that the characters in a fairy tale are not actually conscious." He later points out that characters in fairy tales "seldom have names of their own. More often than not they're known by their occupation or their social position, or by a quirk of their dress: the miller, the princess, the captain, Bearskin, Little Red Riding

---

* It is said that the Grimms valued the sanctity of motherhood above all else and thus vilified the stepmother in the tales they recorded. Ludwig Bechstein, who wrote the Grimms' equally popular contemporary competition, explicitly rejected this, believing that since so many women die in childbirth, Germany had stepmothers aplenty who already had enough problems without always being the villain.

Hood." Objects in the story usually have no description beyond single-word modifiers, such as dark forest, peaceful kingdom, and humble cottage.

But what they have in abundance is action. Things happen. The stories are stripped bare of everything else to make room for nonstop happenings. Raymond Chandler once offered a piece of literary advice: "When in doubt, have a man come through a door with a gun in his hand." Fairy tales live this mantra; they are constant streams of men coming through doors with guns.

But how old are these stories? Until recently, many scholars believed that they were modern creations, perhaps just five hundred years old. Compelling research now suggests that the stories are old. Very old. "Beauty and the Beast" and "Rumpelstiltskin" are about four thousand years old, Jack first climbed that beanstalk five thousand years ago, and a tale called "The Smith and the Devil"—a Bronze Age Faustian tale—goes back six thousand years.

How do we know this? Folklorist Sara Graça da Silva and anthropologist Jamshid J. Tehrani together used a toolkit of evolutionary biology called phylogenetic comparative methods (PCMs) to get at these ages. Biologists use PCMs to group different species by certain heritable characteristics they have in common. By this method, you can infer that they are related and in fact share a common ancestor. Stories are like organisms in that new stories inherit traits from their parent sources. Just as we concluded that the accounts of the Big Dipper with the helpful bird shared a common parent, da Silva and Tehrani applied this sort of reasoning in a structured, mathematical way.

They started with the Aarne-Thompson-Uther (ATU) Index, a list of over two thousand story types. For instance, ATUs 300–749 are tales of magic, with the 300s being a subset called supernatural adversaries; ATU 328 is "The Boy Steals Ogre's Treasure," that is, "Jack and the Beanstalk." Da Silva and Tehrani took the tales of magic, along with stories from two hundred different societies, and set these against the language trees used by linguists to track language families. Stir all that together, add a dash of Bayesian analysis, a handful of Markov chains, and voilà, you get what are probably pretty reliable dates for some stories, with "The Smith and the Devil" being the oldest, predating writing. The technique is quite specific in its findings. Da Silva and Tehrani concluded, for instance, that "Little Red Riding Hood" was created between Europe and the Middle East about two thousand years ago. On a Thursday. Well, okay, not quite that specific.

Fairy tales are manifestations of the culture that birthed them. In "Jack and the Beanstalk," the protagonist is a European who invades a foreign land, kills all the locals, and leaves with the gold. "Beauty and the Beast" is believed to be a tale to inculcate in girls an acceptance of arranged marriages. In many of them, it almost seems as if we are missing some key pieces of backstory. To wit, one must question the true activities of Jack and Jill given that one never goes *up* a hill to fetch water, since water famously flows downhill. And why, for that matter, would it take two of them to get a single pail of water anyway? Some fairy tales may have basis in fact. The "modern" story of the Pied Piper of Hamelin may sound like just another fanciful tale until one learns that in Hamelin in 1384 it was recorded that "it is one hundred years since our children left." A fifteenth-century source adds that it was June 26, 1284, when the piper led away 130 children.

Why are fairy tales so violent? In the original conclusion of "Snow White," for instance, the queen was made to wear red-hot iron shoes and dance for everyone until she was dead. If that was in the 1937 Disney version, I must have been up getting popcorn during that scene. Could it be that people loved stories this gruesome? That's not an unreasonable assumption when set against the backdrop of the billion dollars that the *Saw* franchise has grossed. Perhaps they are that violent because they ironically become less scary for their excess. Wile E. Coyote stubbing his toe may actually cause more of a wince in a viewer than his eating TNT and blowing up. Or maybe fairy tales are violent and scary because they were birthed in a world that was violent and scary. Children really should avoid getting lost in the dark woods or something bad could happen involving wolves or bears.

Author Adam Gidwitz wrote in the essay "In Defense of Real Fairytales" that there are several answers to this question, but one in particular stands out. He writes: "The real Grimm fairy tale takes a child's deepest desires and most complex fears, and it reifies them, physicalizes them, turns them into a narrative. The narrative does not belittle those fears, nor does it simplify them. But it does represent those complex fears and deep desires in a form that is digestible by the child's mind." He goes on to say that the child plays all the roles in the fairy tale, experiences all the action. Gidwitz would explain the absence of specific detail in a given fairy tale as a feature, making the story a blank canvas onto which each child can project themself.

With all of these stories of fear and death, each more macabre than the next, what in the world is Agora thinking about? Gidwitz's quote takes on a new sense when one regards the child he references as the still-nascent Agora. Consider the world at that time: We had adopted agriculture, settled down, and begun forming cities. Our group sizes were substantially bigger as population had grown rapidly. There were now strangers among us. New kinds of dangers had emerged. Privation, disease, poverty, and early death were all around. Agora seems to be confronting a harsh existence in a cruel, or at least indifferent, world. The few happy endings that are actually found in ancient fairy tales are usually set against a backdrop of prior suffering and torment. How different this is from the much smaller and younger Agora contemplating the vastness of the night sky.

Era 3: Civilization and writing (4,000 YA to 500 YA)

Aesop, whose name is indelibly linked to fables, lived about 2,500 years ago. Whether he coined his stories or wrote down those of others is not known. In fact, we know nothing firsthand of the man himself, but his stories are mentioned all over antiquity, by notables such as Aristotle, Herodotus, and Plutarch. Legend has it that he had been a slave, was freed, and later became an adviser to kings. The most unusual thing about our surviving accounts of him is how incredibly ugly he was. But these descriptions date from well after he was purported to have lived, so either they are apocryphal, or the man's appearance was so loathsome that his hideousness was passed down orally for centuries.

In fables, the moral of the story can be explicit or implicit as long as it is so evident as to be screamingly obvious. But sometimes we still manage to miss it. What, for instance, do you remember as the moral of the story of the tortoise and the hare? Perhaps you recall it as a tale of stick-to-itiveness, where "slow and steady wins the race." This interpretation has been contested for literally thousands of years by those who say that the story is clearly about the arrogance of the hare—taking all those naps and not applying his talent—rather than the tenacity of the tortoise. One particularly macabre version of the story has all the woodland creatures dying in a fire because after the tortoise won the race, they all assumed he was the fastest and put him in charge of warning them of danger.

Many proverbs used to be parts of stories until the story entered the collective consciousness to such a degree that the story could be dropped, forgotten even, but the moral would live on. You just have to say three words, "Never cry wolf," to convey the entire message of that tale.

Agora had grown much larger by this time. It wasn't a discrete group of 150 humans anymore. It was a city that perhaps had tens of thousands of people. So Agora had to contemplate a whole new form of ethics, the proper way to conduct oneself in a world with progressively larger groups and higher division of labor. A world with money, wages, and property, each of which were new enough that correct action hadn't all been worked out. That's why these stories are much more practical in day-to-day terms than the earlier fairy tales.

At the same general time as Aesop, but a few countries over, we have Greek myth, which amazingly still permeates our culture. Who doesn't know about Zeus and Hades, Prometheus and Atlas, the Minotaur and Medusa, and Cerberus, the three-headed dog that guards the Underworld, keeping the dead from escaping? They are regularly portrayed in big-budget Hollywood films, TV series, and Broadway musicals. Plays of their exploits written 2,400 years ago are routinely performed around the world, sometimes even in the ruins of the open-air theaters where they first enthralled their audiences. Their myths are updated, retold, and reset in new times and places. Idioms about them live on in our conversations, such as a Herculean task, an Achilles' heel, a Trojan horse, the Midas touch, a Pandora's box, and a Gordian knot.

Why do they have such a hold on us? They survive because they are the founding myths of Western civilization. As Will Storr writes in *The Science of Storytelling*, "Western children are raised in a culture of individualism which was birthed around 2,500 years ago in Ancient Greece. Individualists tend to fetishize personal freedom and perceive the world as being made up of individual pieces and parts. This gives us a set of particular values that strongly influence the stories we tell." He adds that this individualism may have arisen in Greece partly because its hilly and rocky landscape couldn't accommodate large group activities like mass farming.

The structure of the stories often features a protagonist, who goes on an adventure, fights monsters, wins love, earns riches, and returns home to great honor. It's the "hero's journey," and maybe to us it seems like the only way a story can be, but it was once a big new idea: that an individual had the power to choose their own way and wasn't merely a slave or a soldier or the puppet of some god, but the prime mover of their own life. Unlike the fairy tales with the nameless, one-dimensional characters, the characters of Greek myth are named and as multidimensional as you and I.

Storr contrasts this to other cultures, such as ancient China, which "was a realm so other-focused there was practically no real autobiography for two thousand years. When it did finally emerge, life stories were typically told stripped of the subject's voice and opinions and they were positioned not at the centre of their own lives but as a bystander looking in."

Those Hellenistic myths form the basis of our most-loved stories, from *Star Wars* to *The Lord of the Rings*. They are how we imagine our lives—or at least hope them to be. They empowered us to be the prime movers of our own lives. And if that weren't enough, the literary corpus that survives from Greek antiquity is the seed from which all other aspects of Western civilization have grown. Upon Greek culture was built the Roman Empire, with its enduring language and legal code, as well as the Byzantine Empire, which preserved the Greek culture that ultimately gave us both the Renaissance and the Enlightenment.

What was Agora thinking with these stories? As odd as it may seem, I think in the stories of the gods, Agora was embracing humanism. In them, the line between god, demigod, and hero is pretty faint, and all three are really just amped-up humans. But more pointedly, they were humanistic because they showed powerful people who had agency and will. The gods weren't that powerful—Zeus couldn't be everywhere at once—so you were largely on your own to make your way in the world. A mortal man could even, like Heracles, become a god himself.

Era 4: Movable Type (1500 to 1900)

The modern age has brought us entirely new ways to tell stories. In Germany in the mid-1400s, something big happened: Gutenberg's invention proved to be a tipping point for the emergence of widespread, inexpensive printing, unleashing a huge demand for books. Printers started publishing whatever they had close at hand: Bibles, books on theology, and books from antiquity. These were all well and good, but to the average Johann on the street, not very interesting, especially if he didn't know Latin. The public demanded something new, something novel, and thus the novel was born. That really is the origin of the name.

By 1500, printers had cranked out perhaps thirty million copies of around fifty thousand different titles. Examples of about half of those titles survive to this day, and the number of surviving copies of incunabula, the word for books printed before 1500, is around five hundred thousand. About a tenth of them had illustrations.

The next century would see around two hundred thousand new titles coming out, with an average print run of about a thousand copies. For the first time, a person could make a living solely as an author. That two hundred million copies of books could be printed when the literate population of Europe was only around twenty million is a testament to the huge demand for the written word.

These books weren't all just stories, of course, but plenty were. At the beginning of that century, Miguel de Cervantes published *Don Quixote* and

William Shakespeare cranked out a volume of sonnets. His plays would be published as well, but only after his death. Fan fiction was born early: in 1663 Milton wrote a Bible fanfic called *Paradise Lost*.

The novel really was something new. Until then, plots drove stories, and characters were dragged along for the ride. The novel inverted it: complex, nuanced characters with hopes and fears and wants drove the action, and the plot shrank in significance. In her book *Inventing Human Rights*, historian Lynn Hunt credits the novel with bringing about our modern notion of human rights. She writes, "Readers empathized with the characters, especially the heroine or hero, thanks to the workings of the narrative form itself . . . [N]ovels created a sense of equality and empathy through passionate involvement in the narrative."

What do we take away from all this? Clearly, the fact that until this time, Agora's growth was severely limited by its need to use people to remember things. Agora wanted hundreds of thousands of stories, not just a few. Suddenly, a single book could be printed a thousand times, and that storyteller was effectively in a thousand places at once. Since books were often loaned and passed around, a book might well be read by thousands of people.

Superorganisms achieve their superpowers in proportion to the amount of internal communication that is going on. Printed books allowed for exponentially more interactions, even though those interactions are what we would now call *virtual*. Now, one person could talk to thousands over the course of centuries, and the number of interactions between people through books became orders of magnitude higher. Imagine how slow and plodding Agora's thinking was before the advent of movable type. Because of this supercharging of Agora's intellect, we got the Enlightenment, the scientific method, the Industrial Revolution, and the ability to create objects that no single human knew how to make. But Agora knew how to make them.

Era 5: Mass Media (1900 to today)

According to a Nielsen study, Americans spend eleven hours a day consuming media, which Nielsen counts as reading, listening, and watching. Eleven hours. Not all of that is stories, of course, but a big chunk is. How did we go so media crazy? Do we have a limitless appetite for stories? Let's wind back the clock and see how we got here.

Before about 1900, there were two kinds of stories: those told by people in your immediate proximity and those written in books. Stories started being "told" in silent films around 1900. By 1910, there were thirteen thousand theaters in the US. Before 1917, films were shot with an alternate depressing ending to be shown exclusively in Russia. That sounds like a joke, but it isn't. By 1920, the storyteller had entered many homes in the form of a radio. By turning the AM dial, you could switch stories at will. Sound and images merged in talkies, and by 1945, ninety million Americans went to the movies in any given week, out of a total population of just 140 million. People went story crazy, but things had just gotten started. Next came the social juggernaut of television, which exploded in popularity. Today, the average cable subscriber has more than two hundred storytellers—channels—to choose from twenty-four hours a day.

The 1980s brought the personal computer. Suddenly, you could play games that were stories in which you were the protagonist who drove the plot. The 1990s brought the web, and, well, that brought everything else. Millions of stories can be ordered up on demand, for free, on any topic you can imagine.

That's how we got to eleven hours a day of media consumption. How could we not? It all just layers on itself. We still read novels, go to movies, listen to the radio, watch TV, play video games, and use the internet.

The eleven hours will likely go up as we layer on more storytelling formats. Virtual reality promises to immerse us in our stories. You can be one of Beowulf's men or a student at Hogwarts. Neural interfaces may come along that can fool our senses and immerse us even further. You will be able to smell the worlds, taste their foods, touch things in them. Eventually, science fiction speculates, you may live entirely in the story, never leaving it. Some people believe that's where we are right now, living out our entire lives in a world that seems real but is only a computer simulation.

Agora's appetite for stories and other kinds of knowledge still has not been sated. Our technology gives us nonstop feeds—interesting word, feeds—of stories all day long through social media and news sites. Plus, there is still TV, radio, and all the rest. Agora, it seems, wants to know everything.

In these five eras, we can see, at least in outline, the stories that Agora told itself changed as it grew and matured. But as mentioned earlier, we would also expect to see a second type of story, one that never changes, that reflects our unchanging humanity. Do we see this?

Archetypal Stories

E arlier we looked at the ATU classification system, in which thousands of tale types have been meticulously documented and arranged hierarchically. For more than a century, folklorists have been honing and expanding the system, and they are constantly adding new tales as examples of each type. For instance, under "Tales of Magic 300–749" is "Supernatural Helpers 500–559," and number 500 is "Rumpelstiltskin." Who would have imagined that "Rumpelstiltskin" is a general plot, not just a story? The type can be described as "a supernatural helper performs some much-needed task under the condition that a great evil will befall the protagonist if they are unable to determine the helper's name." Although this is the Rumpelstiltskin (German) type, the helper has different names in all the different traditions. These include (spoiler alert) Terrytop (Britain), Whuppity Stoorie (Scotland), Gilitrutt (Iceland), Khlamushka (Russia), Tarandandò (Italy), and dozens more.

Life is so varied that it's no wonder there are thousands of story types, each of which contains many different examples. Or so one might think. But when you boil it all down, are there really thousands of types of stories? Or at their core, are there just a few?

There are, of course, an infinite number of stories. In the 1960s, a semi-documentary TV show about the people of New York would end every episode with the line: "There are eight million stories in the naked city. This has been one of them." But an interesting question emerges, involving just how varied these stories really are. How many different plots are these many stories actually based upon?

The eighteenth-century Italian dramatist Carlo Gozzi once wrote that there were just thirty-six different plots—or dramatic situations. Thirty-six! That's a small number. But according to Johann Goethe, "Gozzi maintained that there can be but thirty-six tragic situations. [Friedrich] Schiller took great pains to find more, but he was unable to find even so many as Gozzi." The nineteenth-century French writer Georges Polti was captivated by Gozzi's claim, writing, "Thirty-six situations only! There is, to me, something tantalizing about the assertion," especially since "he who declared it . . . had himself the most fantastic of imaginations." So Polti dove into the question and in 1895 emerged with an entire book called *The Thirty-Six Dramatic Situations*. Most of them involve tragedies of either family or love, and a sampling of his thirty-six includes *enmity of kin, self-sacrifice for an ideal, conflict with a god*, and *recovery of a lost one*.

So is thirty-six the real number of story types? Not so fast. Ronald B. Tobias's book *20 Master Plots* opens with an apt William Faulkner quote: "If a writer has to rob his mother, he will not hesitate; the 'Ode on a Grecian Urn' is worth any number of old ladies." He warns his readers to be "suspicious of any magic number of plots" and maintains that his book doesn't claim there are just twenty plots but contains "twenty of the most basic plots." His include *quest, underdog, descension*, and the crowd-pleasing *wretched excess*.

But perhaps the twenty have a good deal of overlap, and there are really only seven plots. Christopher Booker spent thirty-four years writing a Jung-inspired book called *The Seven Basic Plots*. His seven are: *overcoming the monster, rags to riches, the quest, voyage and return, comedy, tragedy*, and *rebirth*. Each plot contains exactly five acts, which begin with the protagonist askew in some way and culminate in their achieving perfect balance.

So do we settle on seven plots? Nope. Give the computers a chance to cast their vote. Kurt Vonnegut wrote in his autobiography that he thought his "prettiest contribution" to culture was his 1965 master's thesis in anthropology, which had been unceremoniously rejected by the University of Chicago. He summarizes the main idea as "stories have shapes . . . [and] the shape of a given society's stories is at least as interesting as the shape of its pots or spearheads."

Vonnegut said that stories have a certain shape that he could draw as a curve on a graph. The x-axis is time, and the y-axis shows the experience of the protagonist. The line goes up and down according to the vicissitude of the story arc. He points out, for instance, that Cinderella is fundamentally the same as the story of "the fall" in Genesis. The low point was the ejection from the Garden of Eden (lining up with Cinderella's flight from the ball) to the happy ending of the promise of redemption (and Cinderella's marriage to the prince).

Inspired by this, researchers from the University of Vermont and the University of Adelaide wrote an AI program using sentiment analysis, which maps words and phrases in stories to the sentiments being expressed. They found that most of the 1,700 works of fiction they analyzed fit within one of six categories: *rise, fall, rise then fall, fall then rise, rise then fall then rise,* and, as you have guessed, *fall then rise then fall.*

Perhaps six is still too many. William Foster-Harris's *The Basic Patterns of Plot* argues that there are really just three plots. The first two are *happy ending* and *unhappy ending.* One might reasonably assume that Foster-Harris didn't need a third, given that happy and unhappy would seem to cover the full gamut of possibilities, but no, there is a third one, the *tragedy wherein the main action happens at the beginning and the story consists of the chain of inevitable events that follow.* To Foster-Harris, each of these story types hinges on a single characteristic of a central character: virtue, selfishness, and touched by fate, respectively.

But when all is stripped away, perhaps it turns out there is really just one plot. You probably saw that coming. The one plot is called the monomyth, and it is almost synonymous with Joseph Campbell, author of the 1949 book *The Hero with a Thousand Faces.* Campbell is a giant—arguably *the* giant—in the field of comparative mythology. He says that all stories are really just one story, that of the hero's journey. As he summarizes it, "A hero ventures forth from the world of common day into a region of supernatural wonder: fabulous forces are there encountered and a decisive victory is won: the hero comes back from this mysterious adventure with the power to bestow boons on his fellow man." The hero's journey consists of seventeen stages, starting with the "call to adventure," then the "refusal of the call," "supernatural aid," and so forth. George Lucas famously modeled *Star Wars* on the hero's journey, referring to Campbell as "my Yoda."

Campbell believed that this common pattern underlies most great myths regardless of the place and time of their creation. This is not because it is such a crowd-pleaser of a plotline, but rather, according to Campbell, because in a great cosmic sense, we are all on the hero's journey. Every one of us. That's why it shows up everywhere in all times. Campbell explains our individual quests:

> What is it we are questing for? It is the fulfillment of that which is potential in each of us. Questing for it is not an ego trip; it is an adventure to bring into fulfillment your gift to the world, which is yourself. There is nothing you can do that's more important than being fulfilled.

To Campbell, it is no coincidence that we are all on this identical journey. He believed in the psychic unity of mankind, that everyone shares the same basic mental framework, an idea that had influenced Jung's concept of the collective unconscious. Campbell maintained that this universal journey is hard to have discourse about because "the best things can't be told," and the only way we can access that transcendent reality is through myth. This wasn't a religious belief, per se, for Campbell was quick to say, "I'm not a mystic, in that I don't practice any austerities, and I've never had a mystical experience . . . I'm a scholar, and that's all." When asked what kind of yoga he practiced, he said, "I underline sentences."

There are other ways to classify stories than by their core plot. Aristotle defined story types by the sequence of emotions they elicit in their audience. For instance, good tragedies evoke three specific emotions in a defined order: *pity, fear,* and *catharsis.* The story begins with us, the audience, feeling sympathy for a character. Next, there is a central challenge that makes us feel fear for the character. Finally, we feel catharsis. Regrettably, Aristotle does not define in what sense he means the term, and scholars still debate it, but we need not let that distract us. The important fact is that the emotional experience of the story should be the focus.

Beneath the plots—however many there are—are the words and phrases that make them up. Harvard literature professor Albert Bates Lord studied oral stories and found that at the phrase level, there is much cross-pollination. Over a lifetime of recounting their tales, the storytellers learned certain phrases that rolled off the tongue or were exceptionally evocative, and they would use those exact phrases in a range of stories. In fact, in many oral traditions, the majority of the epic was recycled or repeated material from other stories, like an episode made up of TV clips. In addition to reused phrases, Lord found certain elements that were reused almost universally, such as the "rule of three." There are almost always three of something. Three brothers, three riddles, and so forth. "Goldilocks and the Four Bears" doesn't quite have the same ring to it, does it? And where a three won't work, a seven can often be used in its place. "Snow White and the Six Dwarfs" was never in the running.

But why is all of this true? That's the big question. Why are there relatively few plots? As Jerome Bruner writes in *Actual Minds, Possible Worlds*, "Narrative deals with the vicissitudes of human intentions. And since there are myriad intentions and endless ways for them to run into trouble—or so it would seem—there should be endless kinds of stories. But, surprisingly, this seems not to be the case." Why, for instance, is the theft of fire a worldwide myth so common that it has a Wikipedia entry documenting all known examples, currently at fifteen sourced from every populated continent? There are two possible reasons.

First, the stories might be hardwired into our genes. The fact that there are hundreds of human universals—traits found without exception in all cultures everywhere—suggests that much of our behavior is programmed directly into our DNA, or, at the very least, the behaviors that give rise to these universals are. Regardless, we are not blank slates. Recall the research mentioned earlier that even newborns have expectations of how stories ought to unfold.

Our bias to storify everything was famously demonstrated in 1944 by psychologists Fritz Heider and Marianne Simmel of Smith College in Massachusetts. They created an animated film that shows an interaction between a large triangle, a smaller one, and a dot. It is easy to find online and worth a watch. One cannot help but to see a story in it, of a man and woman walking together when out pops a bigger man who beats up the smaller one, then chases after the woman. Upon coming to, the smaller man pursues them both and then flees with the woman. Heider and Simmel showed this film to thirty-six students and asked them to describe what they saw. Virtually all saw a story about people, and all but one identified the large triangle as a bully.

Where did that story come from? Or, for that matter, the similar stories some people hear in certain pieces of classical music? With a few strokes of a pen, a cartoonist can draw three panels that tell a story, and we effortlessly fill in backstory. Maybe the stories are there in our genes, or they lurk in a collective consciousness in a way powerful enough to compel fifteen unrelated cultures to imagine stories about fire thieves that are so enthralling that they are passed down orally for hundreds of generations.

The second possibility is that there just aren't that many possible plots. Plots are series of causes and effects that form the backbone of a story. A movie may be two hours long, but the plot is probably a paragraph. It is a sterilized sequence of related events in which each event is absolutely essential to preserve the unbroken logic of the chain.

To be compelling, plots must be relatable to most listeners, thus stories must deal with broad issues. But then, they have to be interesting. No one wants to hear the harrowing tale of how you debugged an Excel spreadsheet. This further narrows the range of story options. Next, plots need protagonists, and there are relatively few candidates: ruler, warrior, farmer, child, and so on. Then they need a goal, and there is a limited range of those. But even that isn't enough. "Farmer who is hungry decides to go pick an apple" is relatable, but boring. So there has to be conflict. And the outcome must be uncertain. And so at each level we further constrict the pipeline of good stories.

Think about it: Fire is a pretty big deal in the ancient world, right? Sitting around a campfire, you would wonder how we got such an amazing, mysterious thing. We discovered it? Booooring. It was given to us? Yawn. Ahh! It was stolen. But from whom? A god probably. That adds a good danger element to it. But who stole it? Maybe a hero or a helper animal. Voilà! Are fifteen fire-theft stories all that surprising?

It does seem as though Agora tells many of the same stories over and over throughout the millennia. These are the ones primarily about the struggles of humans and their interactions. They don't really change because we don't really change. That's why even though our daily lives have changed immensely in the past four hundred years, the stories Shakespeare penned back then are as relatable now as when they were new. Today we instantly recognize the evil cunning of Iago, the indecisiveness of Hamlet, the ruthless ambition of Lady Macbeth, and the rashness of Romeo. Four hundred years hence, I suspect Agora will still be thinking about the same stories.

The Twenty Purposes of Told Stories

The primary use of stories is internal, as mental constructs for imagining the future. That's what we use them for every day, every hour, whether we realize it or not. Most of the hundreds of decisions we make daily are based on stories about the future. "If I order the fajitas, a big sizzling platter of goodness will be brought to me, whereas if I order the Cobb salad, I will be miserable, but thank myself tomorrow." Over the millennia we discovered that we could give voice to those stories and, through language, share them with others, and this sharing brought about a cornucopia of new benefits, not just for individuals, but for the future of the superorganism, Agora.

Unlike mental stories, told stories can last indefinitely; they spread and mutate, touching our lives in myriad ways. The fact that they are universal among humans and have not been pruned out of our DNA by natural selection means they must serve some important purpose. What are they for? Lots of things, of course. Many individual stories do double duty or, as the Roman poet Horace put it, contain "words at once both pleasing and helpful to life." Thus, the same story can teach ethics, history, and useful knowledge, and be entertaining in the process.

Twenty may seem like a big number, but keep in mind that there are as many types of stories as there are pasta sauces at the supermarket. "Tale" is probably the best umbrella term for what we think of as stories. Tall tales are untrue. Fairy tales generally have magic or talking animals and the like. Fables have morals. Parables teach a moral lesson usually without animals, magic, or the closing line of "and the moral of this story is . . ." Allegories are stories in which different characters and events represent real-world characters and events. Ballads are stories in the form of poems with short stanzas. Myths are traditional stories about purported true events in the past, often of a magical nature. Legends are myths that could have happened in the world as we understand it. Epics are long historical poems. Sagas are the same, but in prose. And this is only the tip of the iceberg. There are still jokes, plays, dramas, comedies, tragedies, fantasy, and science fiction. In addition, told stories vary immensely in length. The average Zen koan is under a hundred words long, while the ancient Indian epic *Mahabharata* is 1.8 million words, not quite twice the size of the seven Harry Potter books. Thus stories are literary Swiss army knives with, by my count, twenty blades (or purposes). Let's look at them.

To begin with, Agora uses told stories to expand and perpetuate its knowledge. We learn, with hope, from our experiences. However, that wisdom often comes hard and at great cost in the form of bad decisions and suffered consequences. One life is relatively short, and so if you had to learn everything through first-person experience, well, that would be slow going. To top it off, when you died, all your hard-earned wisdom would perish with you, forcing the next generation to start over. But with stories, we can **expand our own knowledge through the experiences of others (purpose number one).** Thus your experience as a human can be thought of as the total of all of your memories of everything you have done *as well as every story you remember.*

As evolutionary psychologist/anthropologist Michelle Scalise Sugiyama puts it, "The opportunity to share verbal recreations of experience greatly increased the amount of knowledge a person could acquire in a lifetime, while greatly reducing the costs of acquiring it." Thus, through stories, it was no longer necessary "for individuals to invest large amounts of time and energy or risk their lives seeking out first-hand educational opportunities."

Along with our ability to plan for the future, our expansive knowledge is the other thing that makes us the dominant creature on the planet. A calf is born knowing how to be a calf; a baby lizard has mastered lizarding moments after hatching. But a human, in our modern society, starts adding knowledge on day one and never really stops adding to their store. The flourishing of the species, perhaps even its survival, is dependent on our preserving what we know and adding to it. Today, of course, we can store it all on paper as well as electronically, but for 98 percent of our history we had neither of those.

We discussed episodic memory earlier. That is your memory of specific events as well as your experience of them. Episodic memory is recalled in the form of stories and is a big part of how we plan for the future, which we conceive of as stories as well. The stories you hear can be stored the same way as actual events that happen to you and are thus available for recall in the same way.

Stories are quite a mental exercise and **hone the brain (two)**. As we engage with stories, our mental facilities are usually on high alert. We are trying to anticipate what will happen and tracking all the assumptions of the story's universe and following the motivations of the characters so carefully that if they do something even slightly out of character, we are pulled back into reality with the thought, "She wouldn't have ever done that." In fact, we even imagine other courses of action for the characters as we say, "Why didn't she just . . . ?" Anticipating what will happen in the story hones the brain such that we can bring the same skill to our real-life social interactions.

Stories are a good way to **teach information (three)** versus, say, a lecture. It turns out that simply listening to someone tell stories about doing something—bird-watching, fishing, or baking—is a more durable form of learning than simply hearing an exposition on those topics, even though the latter is more direct and thorough. This works only for some sorts of skills, though, and is probably not the best way to learn algebra, for instance.

In a similar fashion, stories **act as mnemonic devices for remembering knowledge (four)**. We can tell each other things we have learned. "Always change the batteries in your smoke alarms" is probably a good tip, but how "sticky" is it? On the other hand, the graphic story of your house burning down, leaving you destitute because you didn't change the batteries in your smoke alarms, will—ahem—sear that memory into your mind. The stories Jesus told are more well known than his teachings. Many non-Christians can explain in broad strokes the parables of the prodigal son or the Good Samaritan but likely can't recite much of his theology.

Author Neil Gaiman believes stories are vessels to store boring facts so they will be remembered. In a talk given to the Long Now Foundation, he relays a story from Native Americans in the Pacific Northwest that dates back thousands of years. It's about the forbidden love between a young man and a woman of incredible beauty. Their love was punished by the gods: the ground shook, the sky filled with black snow from a nearby mountain, and then the top of that mountain turned to fire, killing many. The tumult stopped only when the young woman was pushed into the flames of the volcano. Gaiman maintains that the knowledge that volcanoes can emerge and that there are warning signs such as earthquakes would last only about three generations if it were preserved as simple facts. But to make that knowledge last, you must wrap it into a story—one that people love to tell, of angry gods and forbidden love, and somebody being sacrificed to a volcano. Gaiman added, with a visible twinkle in his eye, a tip to would-be storytellers: "People getting thrown into volcanoes—it always works."

Remembering boring facts for a long time is still a challenge. The field of nuclear semiotics is about communicating to the future across "nuclear time." How do we warn people in the distant future of the danger of the waste buried at a disposal site? In 1981, the US Department of Energy and the Bechtel Corporation assembled the Human Interference Task Force, an eclectic team of the intelligentsia, to answer that question. Sure, the normal precautions could be taken with all kinds of warnings, but even if they could be read in ten thousand years, they may have no more effect than the curses written on Egyptian tombs have on us. So, what other options are there? Linguist Thomas Sebeok suggested "that information be launched and artificially passed on into the short-term and long-term future with the supplementary aid of folkloristic devices, in particular a combination of an artificially created and nurtured ritual-and-legend." In other words, make up really good stories about it. To reinforce the stories, he suggested forming an "atomic priesthood" to create annual rites and retell the legends year after year.

Another story-based suggestion, from French author Françoise Bastide and Italian semiotician (specialist in signs and symbols) Paolo Fabbri, was to genetically engineer cats that glow in the presence of radiation. Stay with me here. The idea is that people have kept cats for thousands of years and presumably will continue to do so. We could create a set of stories, folktales, and legends that talk about how if your cat starts to glow, you should leave the area. The podcast *99% Invisible* tasked musician Emperor X to start these legends. As he put it, "I had to write a song about nuclear waste so catchy and annoying that it might be handed down from generation to generation over a span of ten thousand years." The result was his admittedly catchy song "Don't Change Color, Kitty."

Agora also uses stories to support society. One long-standing purpose of storytelling is to **enforce social norms (five)**. Hunter-gatherer societies don't have written law codes, religious texts, or books by Miss Manners. Thus their social norms have to be conveyed a different way, through stories. According to anthropologist Daniel Smith, storytelling can "function as a mechanism to disseminate knowledge by broadcasting social norms." His team studied stories from hunter-gatherer societies in Africa and Asia and found that "of 89 stories, around 70% concerned social behaviour, in terms of food-sharing, marriage, hunting and interactions with in-laws or members of other groups."

In *The Storytelling Animal*, Jonathan Gottschall adds to this thought, saying that story "continues to fulfill its ancient function of binding society by reinforcing a set of common values and strengthening the ties of common culture. Story enculturates the youth. It defines the people. It tells us what is laudable and what is contemptible."

Often, socially acceptable behavior is illustrated using the trickster character. The trickster can be a human, a deity, or an animal with bad PR, like the crafty fox or the messy pig. They can either embody a single bad characteristic that is the subject of the story, or they can embody the seven deadly sins plus a few more. The trickster is a boor; he mooches off others, never does his share of the work, and—the ultimate sin in a forager society—always thinks of himself before the group. The ultimate failure of the trickster in the story and their corresponding misery contains an obvious message: If you do this, the same fate will befall you. Subtlety is not a trademark of these stories, lest the message be lost.

Dealing with real-life free riders in forager societies was a delicate matter. The people in your group were your safety net. In the book *Debt*, anthropologist David Graeber relays the account of a Maori named Tei Reinga, a lazy glutton who was always asking fishermen "for the best portions of their catch. Since to refuse a direct request for food was effectively impossible, they would dutifully turn it over; until one day, people decided enough was enough and killed him." That story was probably retold many times in the presence of other freeloaders.

In contrast to the "hit them over the head with the moral" approach, there is a subtle way that stories are used to **gently shape behavior (six)**. Stories can be told in a way that listeners are allowed to come to their own conclusions, that is, apply the story to their lives as they see fit. This is common in religious traditions. Jesus taught in parables that required reflection by the hearers. Even his disciples didn't always understand them and would ask for explanation. This manner of teaching had been used five hundred years earlier by both Buddha and Confucius, and undoubtedly centuries before that.

An episode of *Star Trek: Voyager* illustrates this use of stories. A member of the crew, B'Elanna Torres, has crash-landed on a planet that is exactly like ancient Greece, except of course the people speak English with American accents. It's best not to think too hard about that part. Anyway, she finds herself in a country that is on the brink of war, and she befriends a young peace-loving playwright named Kelis who is resolved to write a play to show that war is not the answer to his leader's disagreements with their neighbor. The two characters have this exchange:

Kelis: My patron is filled with hatred for his rival, so our play should be filled with love.

Torres: You can't change somebody's way of life with a few lines of dialogue.

Kelis: Yes, you can. It's been done before . . . Why can't my play take the place of a war?

Of course, the plan works. The ruler sees the protagonist throw his weapon away and opt for peace, and it enables him to see that path as well.

Melanie Green is a social psychologist at University of Buffalo who researches the power of stories to change beliefs. She believes that when we hear or imagine stories, we are transported into that narrative world and our brains get fully engaged with it, which means our range of emotions come to bear on it. As she says on the podcast *You Are Not So Smart*, when that happens, "we have this in-depth and this vivid experience, that we take the information from the stories and bring it into the real world with us. So we learn from what happens to the characters in the stories, and that can change the way we think and do things."

She adds an interesting tidbit: "We've had the consistent finding in my work that it doesn't really matter whether you tell somebody that a story's fact or fiction . . . [T]hey're persuaded by both of them." That's fascinating. Even if they know the story is false, it affects them the same way. This is similar to how in medicine you can still observe the placebo effect even if you tell people they are getting the placebo.

Another use of stories is to **teach personal ethics (seven)**, often to children. Frequently, this is done by example, not through explicit moralizing. In these tales, there is good and evil. Despite setbacks along the way, good wins out in the end, and the hero vanquishes the villain. Bad deeds are punished, not necessarily by authorities, but by the intrinsic nature of the deeds themselves. Likewise, virtue is rewarded. Tell a thousand of those kinds of stories to kids and those basic patterns are laid down in their psyche. This idea is at the core of the second book in C. S. Lewis's Space Trilogy, *Perelandra*, which takes place on Venus. The story of Eve's temptation by Satan is set up again. In it, God has given Venetian Adam and Eve just one rule, then Satan manages to get Eve alone and starts tempting her. The way he does this is interesting. He never tells her she should break the one rule. Instead, he just tells her stories, day after day, about women who were given a good rule, but a situation arose where the right thing to do was to break the rule. And knowing when to keep the rule and when to break it was the essence of wisdom. In all his stories, the rule-breaking woman receives acclaim after the fact for doing the smart thing.

Stories support society in that they both **create culture (eight)** and **act as social glue (nine)**. In his book *Parallel Myths*, J. F. Bierlein writes, "Myth is the 'glue' that holds societies together; it is the basis of identity for communities, tribes, and nations." Jonathan Gottschall, mentioned earlier, echoes this thought, calling story "the grease and glue of society: by encouraging us to behave well, story reduces social friction while uniting people around common values." Even today, the founding myths of a country, along with its storied history, still connect all the people living there. Joseph Campbell pushes this further, writing that "the rise and fall of civilizations in the long, broad course of history can be seen to have been largely a function of the integrity and cogency of their supporting canons of myth; for not authority but aspiration is the motivator, builder, and transformer of civilization."

Stories in popular culture bind us together as well. Three times as many people can name two of Snow White's seven dwarfs as can name two of the nine US Supreme Court justices. Twice as many can name the Three Stooges as can name the three branches of the US government. Nearly twice as many people can name Superman's home planet as can name the planet closest to the sun. Such facts are often used to mock the average Joe, but seldom is it pointed out how striking it is that these fictional stories transcend ages, races, and economic status. You can say people should know more about civics if you like, but that doesn't mean they should know less Stooge.

At the micro level, stories serve this same purpose. Storytelling itself is by definition a social act. There must be tellers and listeners. In most cases, there are multiple tellers. Whether around a campfire or a dinner table, the idiom of "swapping" stories is an apt one. They are traded for other stories. This is speculation, but this use of stories is probably quite ancient. Imagine you lived twenty-five thousand years ago and you bumped into someone from your group of 150 or so people. What do you talk about? You probably share useful information in the form of stories. "This morning I was walking back by the mountain when I saw two bears . . ."

Stories can be used to **promote empathy (ten)**. In the 1960s, Jean Briggs, a Harvard graduate student, went to live above the Arctic Circle with the Inuit. Her experiences yielded the book *Never in Anger: Portrait of an Eskimo Family*. She observed that the Inuit never showed anger. She connected it to how they use stories to discipline their children. There is no scolding, no yelling, no time-outs. Instead, the parent would wait until the child calmed down and then act out a story. Say you have a child who is biting other kids. The parent might ask the child to bite her, which raises all kinds of questions in the child's mind, such as, "Why would I bite my mom?" If the child went ahead and did it, the mom would recoil in pain, saying, "That really hurt." The same effect could be achieved with two stuffed animals, one of whom bites the other. The parent might ask, "Do you think that bite hurt? Was it nice? Why did they do that?" The idea is to use the story to work through the emotional implications with the child. Yell at the child or scold them, the thinking goes, you just teach them to be angry children, that anger is a proper response. But show them the implications of their actions through story, and they learn empathy.

A number of studies show that consuming fiction increases our ability to discern motives. That's an intellectual thing. But when we are immersed in a story, it also becomes an emotional experience, and we empathize with the characters. Henry David Thoreau once asked, "Could a greater miracle take place than for us to look through each other's eyes for an instant?" That's what stories are.

The power of emotions is that we act on them. For better or worse, they drive us. Knowing something can be useful, but without some kind of passion under it, people are just biological libraries. Empathy is the mechanism through which one individual feels emotions because of the situation of another. Author David Robson writes for the BBC that "brain scans have shown that reading or hearing stories activates various areas of the cortex that are known to be involved in social and emotional processing, and the more people read fiction, the easier they find it to empathise with other people."

There are those who make good arguments against empathy. Paul Bloom, writing for the *New York Times*, is no fan of empathy. He argues that compassion, "caring for others without feeling their pain," is better, channeling his inner Spock to write, "A good policy maker makes decisions using reason, aspiring toward the sort of fairness and impartiality empathy doesn't provide." You will likely agree or disagree with this in accordance with your views on Vulcans. Too much empathy can lead to empathy fatigue, a condition seen in professions such as healthcare where the day-to-day job can be heart-wrenching.

The final way that stories support society is that they are a major mechanism by which we **persuade others (eleven)**. As author Andy Goodman writes, "Numbers numb, jargon jars, and nobody ever marched on Washington because of a pie chart. If you want to connect with your audience, tell them a story."

Governments use stories to persuade, to unite people with a common vision. When we disagree with the vision, we call the stories propaganda. Atrocity propaganda is a kind of storytelling designed to spark outrage. In 1990, after the Iraqi invasion of Kuwait, a girl who provided only her first name, Nayirah, gave highly emotional testimony before Congress that she had personally seen Iraqi soldiers steal incubators out of Kuwaiti hospitals, leaving the babies that had been in them on the floor to die. Only later was it all revealed to be false. Her appearance was part of a sophisticated pro-Kuwait PR campaign, and she was revealed to be none other than the daughter of the Kuwaiti ambassador to the US. In World War I, virtually all belligerents engaged in this sort of storytelling. For instance, German soldiers, it was falsely claimed, would cut off the hands of Belgian babies, or eat them, or even crucify people. In World War II, the Nazis told the German people a story of their own involving Jews, the betrayals that cost them World War I—and Germany's ultimate destiny.

Another way stories are used to persuade is in commerce. Author and marketing guru Seth Godin says, "Marketing is no longer about the stuff that you make, but about the stories you tell." This has been definitively shown via an experiment by writers Rob Walker and Joshua Glenn called Significant Objects. They bought a bunch of assorted objects at thrift stores for the average price of $1.25. They then had writers pen stories about each one. The objects were then posted on eBay along with their respective stories. The stories were not represented to be true, and they even carried the writer's byline. The nearly two hundred objects they sold brought in about $8,000—an average price of $40 per item.

In the golden age of oversized magazines, the most iconic ads were stories. What is often thought to be the best print ad ever made was one that David Ogilvy wrote for Rolls-Royce, a company with an ad budget that was just 2 percent of Cadillac's. Ogilvy said he was asked to perform a miracle, to write ad copy that everyone would read—and never forget. The headline he came up with read, "At 60 miles an hour the loudest noise in this new Rolls-Royce comes from the electric clock." The ad went on to tell a story, the story of how carefully every Rolls-Royce is made. Ogilvy later said, "When I presented this headline to the senior Rolls-Royce executive in New York, that austere British engineer said, 'We really must do something to improve our clock.'" The ad was so successful that sales went up 50 percent in one year, and Ogilvy hit his mark: my friend Jason told me about this ad in high school, and thirty-five years later I still remember it. Another Ogilvy classic also carried a story. It was "the Man in the Hathaway Shirt," showing a dapper chap with an unexplained eye patch that became virtually synonymous with that brand. The ad ended with "Hathaway shirts are made by a small company of dedicated craftsmen in the little town of Waterville, Maine. They have been at it, man and boy, for one hundred and fifteen years." You would think they would eventually let them retire.

Agora uses told stories to aid individuals in the superorganism. Stories, for instance, **help us achieve spiritual enlightenment (twelve)**. Multiple traditions have stories upon which the faithful are encouraged to meditate. The stories themselves can be so powerful that they induce religious feelings. Joseph Campbell believed that they are the only path to what he called transcendent reality, which can't be accessed directly with words or logic.

He wrote that "the first function of mythology is to reconcile waking consciousness to the *mysterium tremendum et fascinans* of this universe as it is."*

Stories **are therapeutic (thirteen)**. Not only can they change your mental state, but there is an entire field called narrative therapy whose maxim is "the person is not the problem; the problem is the problem." It helps people identify the stories about their lives they are defining themselves by and then, if they wish, change those stories.

Next, stories **provide escapism (fourteen)** from a harsh reality. That may sound negative, but Neil Gaiman defends it: "Once you've escaped, once you come back, the world is not the same as when you left it. You come back to it with skills, weapons, knowledge you didn't have before. Then you are better equipped to deal with your current reality."

In the 1939 essay "On Fairy-Stories," J. R. R. Tolkien also wrote in defense of escapism, offering an analogy: "Why should a man be scorned if, finding himself in prison, he tries to get out and go home? Or if, when he cannot do so, he thinks and talks about other topics than jailers and prison-walls?" Years later, C. S. Lewis related that the people who are most preoccupied with and hostile to the idea of escape were in fact the jailers.

Finally, and perhaps most obviously, stories **provide entertainment (fifteen)**. Richard Wiseman, a professor at the University of Hertfordshire, joined forces with the British Science Association to perform a "scientific search for the world's funniest joke." After having 350,000 people from seventy countries rate forty thousand jokes, they declared a winner. It goes like this:

> Two hunters are out in the woods when one of them collapses. He doesn't seem to be breathing and his eyes are glazed. The other guy whips out his phone and calls the emergency services. He gasps: "My friend is dead! What can I do?" The operator says, "Calm down. I can help. First, let's make sure he's dead." There is a silence, then a shot is heard. Back on the phone, the guy says, "Okay, now what?"

---

* *Mysterium tremendum et fascinans* is a mystery before which we both tremble and are fascinated by, both repulsed and attracted.

A paradox about entertaining stories is that they often depict situations or events that would not be entertaining in real life. In the *National Lampoon's Vacation* movies, we derive pleasure from watching things happen to the Griswold family that we would never want to have happen to us. Likewise, I recall a *Bloom County* comic strip where the characters are watching TV. They go back and forth trying to figure out whether they are watching a news report about the conflict in Lebanon or an old rerun of *Rat Patrol*. The final panel has one of the characters yelling, "Will someone please tell me if I should be enjoying this or not."

Generally speaking, we want our lives to be somewhat boring. For instance, you never want a Wikipedia entry written about your plane flight. But we delight in the mishaps of others. Weren't stories supposed to foster empathy? That almost seems cruel. Further, why do some people enjoy sad stories that make them cry? Or, even more inexplicable, why do they pay good money to be terrified by movies? Why does the *Saw* franchise even exist? In a similar vein, Willie Nelson said, "Ninety-nine percent of the world's lovers are not with their first choice. That's what makes the jukebox play." In other words, we put quarters in that slot in order to experience melancholy. Why?

University of Pennsylvania psychologist Paul Rozin has a theory. He coined the term "benign masochism" to describe a range of negative emotions and sensations that we actually enjoy if we are 100 percent safe in doing so. These include muscle aches after a hard workout, bitter liquors, stinky cheese, and sad or scary movies. We enjoy our body's reaction to these sensations when we are certain no harm will come to us. We can have the rich experience of revulsion, terror, or sadness because we know it is not real. Joseph Campbell thought something similar to this, saying, "I think that what we're seeking is an experience of being alive."

Agora uses stories in ways that transcend time, that connect the future and the past. It does this in a variety of ways. For instance, we use stories to **coordinate action (sixteen)**. "You do this, and I'll do that, and then this will happen" is a story about the future that probably gets told millions of times a day.

Next, they **explain how things got the way they are (seventeen)**. Octopi are amazing animals. They are incredibly intelligent in spite of being solitary creatures with short life spans, often just a couple of years or so. If they lived longer and in large groups, they might give us a run for our money as top species on the planet. What interesting creatures they are: They keep the majority of their neurons in their tentacles, so they are big, distributed brains. They can change the color of their skin to blend perfectly with their surroundings and yet are themselves colorblind. Their blood is blue, based on copper, not iron. They are almost like an alien species. Why is this? Well, according the Kumulipo, the creation chant of the Hawaiian religion, the universe has been repeatedly destroyed, but each time it is rebuilt anew. The *he'e*—octopus—is the lone survivor from the previous universe who managed to squeeze its gelatinous body through the narrow gap between the worlds.

Another story that explains the octopus is that billions of years ago, complex life emerged on Earth, and through countless cycles of reproduction, favorable traits were passed from parent to offspring. Over time, these traits diverged and became more specialized and suited to their individual environments. Eventually, all the animals of the world emerged, each filling a unique function in an inconceivably vast, interconnected, and interdependent web of life.

They're both pretty good stories, and both serve the same function, to explain the world and tell us something about how it operates. I pair them together not to disparage either, but to bring home a point made by J. F. Bierlein in *Parallel Myths*, where he writes that myth is "the first fumbling attempt to explain *how* things happen, the ancestor of science. It is also the attempt to explain *why* things happen, the sphere of religion and philosophy. It is a history of *pre*history, telling us what might have happened before written history."

Next, stories **preserve history (eighteen)**, giving Agora a past with meaning. Before the written word, how did people know what happened in the past beyond what they would remember? We take knowing history for granted, but just think how untethered that must have felt. "Who are we?"

"How did we get here?" Those sorts of questions were answered through stories. Since those stories were, literally, the only shows in town, they must have had rapt audiences. Throughout their lives, people would hear the same stories, again and again, and in turn would pass them down.

Doug Hegdahl was a sailor in the US Navy during the Vietnam War who was captured and sent to the infamous Hanoi Hilton prison camp. He was to be granted early release, so with the help of another prisoner, he memorized the names of the 256 other prisoners, along with when and how they were captured and other personal information, by setting it all to the tune of "Old MacDonald Had a Farm." He can still recall it half a century later. That must be how long histories were passed down through the ages with fidelity, and why it is entirely possible that the Aboriginal peoples of Australia have a living memory of a volcano eruption thirty-seven thousand years ago, in the form of their myth of a giant turning his body into the mountain Budj Bim, and his teeth turning into molten lava.

In early times, people were living libraries. Storytellers functioned more as preservers of knowledge than stage performers. They kept the past alive. As political theorist Hannah Arendt put it, "Even Achilles remains dependent upon the storyteller . . . without whom everything he did remains futile." They served a critical civil function: the past may be dead, but our stories *about the past* are alive and have real impact. A society without a history is just a bunch of people. July 4, 1776, and September 11, 2001, are not just dates, but powerful stories. The Mayans totally got this. When they conquered a city, they broke and mutilated the fingers of the scribes, ending their ability to record anything, in effect ending that city's history. Individuals can also have their history erased. An ancient punishment called *damnatio memoriae** was practiced across many ancient civilizations, including the Egyptians, Greeks, and Romans. It consisted of removing all references to a person, the record of everything they did, so that it was as if they never existed. Talk about cancel culture.

---

* Literally, condemnation of memory.

Likewise, that's why tyrants rewrite history books, because while they cannot change the past, they can alter the story about the past to try to change the present. In George Orwell's *1984*, Winston Smith works at the Ministry of Truth, where he constantly rewrites history to match the ever-changing state version. Regrettably, this fictional account is all too real in the modern world, where technology often amplifies revisionist histories.

To preliterate people, myth was history, science, and religion all rolled up into one. The modern mind doesn't work that way. Take the question, "What are you made of?" A biologist will say "mostly water," a chemist will say "mostly oxygen," and a physicist will say "almost entirely empty space." Back in the day there was one answer, and it was contained in the myth. French anthropologist Claude Lévi-Strauss wrote that today, "history has replaced mythology and fulfills the same function," which is to make sure that "the future will remain faithful to the present and to the past."

Next, Agora uses stories to **imagine the future (nineteen)**. The theme of this entire section is how we use mental stories to imagine the future as we make plans. But of course, we use told stories to picture it as well. The list of things we once imagined we might have in the future that now actually exist is impressive. Captain Kirk had a flip phone, Uhura had a Bluetooth device in her ear, and there was a $10 million XPRIZE to make Dr. McCoy's tricorder. Ask Alexa what it wants to be when it grows up and it will reply, "I want to be the computer from *Star Trek*."

Apple claimed it invented the tablet, and in a court case where it accused Samsung of infringement, Samsung entered into evidence photos from *2001: A Space Odyssey* that clearly show something that looked like an iPad but predated it by forty years. The production notes of that film revealed that Stanley Kubrick and his team envisioned more functionality for the device than made it into the movie.

It isn't just gadgets and science fiction. Utopian literature across centuries has imagined future Earth. Five hundred years ago, Sir Thomas More wrote a book imagining a future world with religious freedom. A century after that, *Civitas Solis* described a future with no slavery. Another century elapsed, and *The Adventures of Telemachus* envisioned a future with constitutional governments. The nineteenth century was full of books describing futures with legal equality between men and women, universal public education, free preventive healthcare, and a government safety net for the poorest people.

Finally, stories are **used as cautionary tales (twenty)** about what can go wrong in the future. Personally, I am not a fan of dystopian movies. Look, I get it, I'll pay ten dollars to watch some A-lister fight a robot-gone-rogue. After all, who wants to see the movie *Everything Is Wonderful in the Future*? But when I watch those movies, I always think, "That will never happen." I am, unabashedly, optimistic about the future of humanity. I'm a sucker for science fiction, though, so I usually go, eat my popcorn, and roll my eyes a whole lot. But then I stumbled across a quote by the legendary science fiction author Frank Herbert, who said, "The function of science fiction is not always to predict the future but sometimes to prevent it." In other words, the stories are cautionary tales of what could go wrong.

Cautionary tales do lend themselves to ridicule. In 1907 Hilaire Belloc wrote a book of satire called *Cautionary Tales for Children*, which contained stories including "Rebecca: Who Slammed Doors For Fun and Perished Miserably" and "Lord Lundy: Who was too Freely Moved to Tears, and thereby ruined his Political Career."

In Closing

With language, individual humans gained a superpower, but the unintended consequence was that a new kind of superorganism was created, a tribe of humans, Agora. Our conversations were Agora's thoughts, our stories were its wisdom. Our corpus of those stories grew year by year, expanding Agora's abilities. However, the only place Agora had to store information was in our individual brains. We weren't the best data storage devices, but we were all Agora had on hand at the time.

Over the years, the population of humans increased, and we grew ever more specialized. While other kinds of superorganisms had just a few different jobs for their members, Agora had thousands. As such, Agora grew ever smarter. We moved into cities, and Agora grew more powerful. Then we invented writing, enabling Agora's knowledge to grow indefinitely. With this innovation, one generation could build on the learnings of the prior ones. This eventually led to science and modernity.

Even today, our primary use of stories is still mental. We use them to plan for the future, from the next few minutes to years or even decades. But that raises an interesting question: How do we know what will happen in the future? Sure, we have an ability to *imagine* what might happen, but how do you *know* what actually will happen? As Aristotle wrote, "Nobody can narrate what has not yet happened."

Or can they?

In the scientific age, we learned that the world behaved in predictable ways, that it was governed by laws, and those laws could be discerned and mastered. We began to wonder: "Is the future likewise governed by laws? If so, can we learn those laws and use them to predict what will happen?"

That brings us to Act II, set in France in 1654, where two great mathematicians began building such a science, which they called *probabilité*.

ACT II

Dice

The Foreseeable Future

M ental time travel is a human superpower. Our first uses of it must have been relatively mundane, predicting what would happen in the next few minutes. But getting a taste of it, we inevitably wanted more, to see further and clearer. So, we constructed—or discovered—other ways to get a glimpse of tomorrow. How audacious is that? Why in the world *should* we be able to see the future at all? It is the future; it hasn't happened yet. Say what you will about us, we are an ambitious lot.

To predict the future, you have to understand why it happens the way it does. Why do things unfold a certain way and not another? Broadly speaking, there have been four different lines of thinking that attempt to get at the root of that question. Few of us are dogmatic purists on the question, and even if we espouse one viewpoint over the other at an intellectual level, we all live our lives in a pluralistic fashion, blending them all in our daily lives and decision-making.

The first idea is that the future happens as it does because it is *fated to happen* that way. In the words of William Blake, "Some are born to sweet delight. Some are born to endless night." In ancient Greece, fate was the highest power of them all, even over the gods. Your fate was determined by the Moirai, three sisters named Clotho, Lachesis, and Atropos, who plotted out your entire life the moment you were born, like an overbearing grandmother. Clotho would spin out the thread that would be your life, Lachesis would set your destiny, and Atropos would use her scissors to snip the thread, setting the day of your death. You could not escape this fate, but neither could the gods escape theirs. In Homer's *Iliad*, Zeus is unable to save his beloved son Sarpedon, who had been fated to be killed by Patroclus.

In spite of being governed by fate, people still had free will. When the king of Thebes found out that his son Oedipus would kill him, he took a series of prudent steps to prevent it. The king's actions were his own, but he couldn't change his fate. It's like this: Imagine you sit down to play chess with the greatest player alive. Your fate is set—you are going to lose. You are free to make whatever moves you want, but it won't change the outcome. As Plato put it, "Nobody can escape destiny."

The Moirai were the ones who *set* fate. Other gods and oracles could *see* the future but couldn't change it. Apollo knew what was going to happen in the future, and he gave that gift to the oracle at his temple at Delphi, with two instructions: always tell the truth, and never be too specific. Oracles at Delphi performed this duty for around 1,500 years. Many ancients, including Plato, attested to their accuracy. We even know the final message of the final oracle. It was delivered to Julian the Apostate around the year 360 AD. Julian wanted to restore the old gods to their place of preeminence in Roman culture, so he sent a messenger to Delphi. The oracle threw cold water on Julian's plans, informing the messenger that "the speaking water has been silenced."

The Romans adopted this Greek belief about fate, renaming the Moirai as the Parcae. In Norse mythology there was something similar, beings called the Norns, and there were many of them, some good and some bad, which explained why people's fates were so varied. Life's pretty rough if you end up with a mean Norn.

In antiquity, the future unfolded not only according to fate but also by other divine forces—including luck and fortune, both believed to be real, not simply metaphors—which, like fate, doled out their favors and curses seemingly arbitrarily, with little regard for individual merit or piety.

The next belief regarding how the future unfolds is known as *necessity*, also called determinism. The future happens a certain way because it *must*. It isn't *fated* to happen that way—nobody decided it—but rather it can only happen that way. Event A causes B, which causes C, which inevitably leads to D. Everything that happens is the inevitable outcome of all of the events leading up to it, meaning that you stubbing your toe this morning on your bed frame was the inevitable outcome of a series of events stretching back to the Big Bang.

This belief about the future is actually a belief about the nature of the universe; that old debate between monism and dualism. Monism is a belief that there is only one sort of thing, the material, mechanistic, cause-and-effect world that we perceive around us. Democritus, living 2,500 years ago, said it simply: "Only the atoms and the void are real." His teacher, Leucippus, had explained the significance of this: "Nothing happens at random; everything happens out of reason and by necessity." Monism stands in contrast to dualism, which acknowledges the physical world but asserts that there is something additional that is equally real but is immaterial, such as the mind. Plato held this view.

The early Enlightenment thinkers euphorically embraced the monist view of the universe. The world was seen as a giant clockwork, where everything could be explained in straightforward mechanistic terms. *Everything*, even us. As Thomas Hobbes wrote, "What is the heart but a spring, and the nerves but so many strings, and the joints but so many wheels, giving motion to the whole body?" Hobbes saw humans as something akin to simple computer programs, coded simply to avoid pain and pursue pleasure. This was an intoxicating view because it needed no god, only reason, and appealed to nothing but a physical reality that was both knowable and unchanging. Certainly, there was still the mind-body problem articulated by René Descartes, which said that it's really hard to imagine what an "idea" is made of or the mechanics behind free will, but this was seen as a minor objection, and that at the very least, the *physical world* was clearly just a machine.

They had good reasons to believe this. In a *Calvin and Hobbes* strip from 1989, Calvin asks his dad how light bulbs work and his dad replies, "Magic."

Calvin points out that he had said that about the vacuum cleaner as well and speculates that his dad really doesn't know the answer. In medieval times, the answer "magic" was about all we could muster for a wide range of questions about natural phenomena. Then, as we entered the scientific age, we learned how to answer more and more of those questions by discerning law-like behaviors in nature that could be expressed mathematically and make astonishingly accurate predictions. Suddenly it seemed that maybe all of nature was governed by impersonal, knowable laws.

This shift in mindset from magic to science can almost be dated to a literal moment in time: 9:33 AM GMT on Friday, April 22, 1715, as reckoned by the Julian calendar. This is when London plunged into darkness as a total solar eclipse passed through Great Britain. We call this Halley's Eclipse, after Edmond Halley, and it had been widely predicted, with maps of the eclipse's path being published in all the widely read broadsheets of the day. Halley's calculation was off by four minutes and a few miles, but no one was petty enough to quibble with that. Science had said that the moon would block the sun that morning, and it did, and there were millions of eyewitnesses to the event.

To the Enlightenment thinkers, monism had two inescapable implications. First, it meant that the future was predictable. It could be divined through a set of immutable physical laws. Pierre-Simon Laplace, an eighteenth-century polymath, suggested that if there was an entity who knew the present state of everything in the universe, then "for such an intelligence nothing would be uncertain, and the future, like the past, would be open to its eyes." And second, as mentioned previously, it eliminated the need for, even the possibility of, supernatural explanations to account for anything. It was Laplace who, when asked by Napoleon why there was no mention of the Creator in his writings, is said to have replied, "I have no need of that hypothesis." Even God could be reduced to a scientific hypothesis, and an unneeded one at that.

The next idea about how the future comes to be is *synchronicity*. This is a bit difficult to understand because it runs contrary to the logical cause-and-effect deterministic world. In synchronicity, things are related to other things in which there is no causal relation. Although the term synchronicity was coined by Carl Jung, the concept itself is quite ancient.

Consider astrology. Have you ever wondered what is believed to be the mechanism that makes it work? *Why* does it work? I mean, you could take a scientific view and say that people born at certain times of the year gestated though different seasons with their mothers eating different foods, which could perhaps have predictable effects on their fetal development. Thus a Leo might tend to act one way while a Libra another way. In fact, that might even be true—but no one says that. They just say that because you were born at this exact time, at this latitude, with Mars in this position, you have certain characteristics. *Why?* It seems to defy all reason, and yet astrology is the most common form of divination practiced today. There are more astrologers than astronomers, by a wide margin, both in the US and around the world. In China, the fifth most common search term on the Baidu search engine is "astrological sign," and it is not unheard of for job postings to specify preferred signs of applicants. In India, Vedic astrology is used to arrange marriages and decide propitious times to start a business. In the US, the National Science Foundation discovered that a majority of young adults believe that astrology is a science. In addition to being pervasive, astrology is old—incredibly old, if the stars painted on the walls of caves twenty thousand years ago are a form of it, as some believe. In any event, it is certainly older than five thousand years, because we see references to it among our first specimens of writing. In the biblical account of the birth of Jesus, it's interesting that the wise men are from the *east*, saw a star in the *east*, but then traveled *west*—the opposite direction of where they saw the star. Some scholars believe they were astrologers who saw a sign in the sky that the Hebrew Messiah was born, so they headed to Jerusalem, asking King Herod, "Where is the one who has been born king of the Jews? We saw his star when it rose and have come to worship him."

And it isn't just the ignorant or uneducated who believe in it. Enlightenment men of learning still regularly believed in astrology. Tycho Brahe believed that "human beings are influenced by heaven in a lesser degree than animals." Galileo Galilei didn't just believe in astrology; he taught it and worked up horoscopes for patrons and students. Francis Bacon believed in parts of it but argued for reform in how it was practiced. Johannes Kepler wrote a book about it, and eight hundred of the horoscopes he drew up still survive. When he was just twenty-two, he issued a calendar for 1595 that predicted a peasant uprising, a Turkish invasion, and a bitterly cold winter. All three came to pass, bringing him fame.

Astrology and astronomy used to be the same thing. Attempts by early people to make practical calendars by studying the stars served an additional purpose—to infer the will of the gods. The fact that the outcome was frequently incorrect didn't deter belief at all, any more than incorrect weather forecasts make us believe that meteorology is mere superstition. But by a century or so after these men lived, science had almost universally rejected astrology.

But up until then, these really smart genius types truly believed, to greater and lesser extents, in things like astrology. Why? Why in the world did they think it could possibly work? Synchronicity. The stars don't have any causal relationship to you, but they are meaningfully related to you. You are in sympathy with them.

Astrology is by no means the only example of synchronicity in action. All forms of divination rely on it. The Wikipedia entry on divination methods has over three hundred different named ones, words you would totally challenge in a Scrabble game. A few selections from the *a*'s are abacomancy, acultomancy, aeromancy, alomancy, amniomancy, anthomancy, arachnomancy, and apantomancy. These involve reading, respectively, dust, needles, weather, salt, placentas, flowers, spiders, and chance encounters with animals. Practitioners would not find the plentitude of these undermining to their confidence in any of them. If everything is connected in a sympathetic way, then you could divine with anything—say, the underarm sweat stain on a gym shirt—as long as you know how to read it.

Those who read such signs probably weren't frauds. We have found too many clay models of livers covered in notes for interpretation as well as complete texts describing technique, which have much nuance as to the meaning of the size, color, density, texture, and shape of the lobes. Ancient people reborn today would see no real difference between that practice and our aforementioned weather forecasts divined by interpreting squiggly lines on a map.

Another use of divination to predict the future is sortilege, divination by mechanical means, such as casting lots, dice, coins, or other items. One widely practiced method was odds and evens, in which a question can be posed, then a handful of something—beans, grain, or pebbles—is dumped out and counted. An odd number means one thing, even means another. As F. N. David writes in *Games, Gods and Gambling*, "The method clearly does not give the god much scope for self-expression but at least it produces an unequivocal answer."

Archaeologist Laurence Waddell spent time in Tibet in the 1890s and writes of a divination method the Tibetan monks used to predict the form in which you would be reincarnated. It involved a six-sided die and a board. He secured one of these sets and remarked that his die "seems to have been loaded so as to show up the letter Y, which predicts a ghostly existence, and thus necessitates the performance of many expensive rites to counteract so undesirable a fate."

Belief in this sort of interconnectedness, along with oracles, astrology, and all the rest, was predominant through most of history. But it was not universal, even in antiquity. Cicero wrote a two-volume book on the topic called *On Divination*. The first part has his real-life brother, Quintus, making a case for divination, while the second volume has Cicero rebutting it. Many think Cicero was not a believer in divination, but the arguments on both sides are well formed, neither appearing to be set up as a straw man. The book deliberately comes to no conclusion, leaving it up to the reader, though this may have just been good politics for Cicero.

Haven't thinking people left all that behind? Isn't the belief in synchronicity contrary to science? Not necessarily. The mechanism of the universe—and thus how the future unfolds—may well be more complicated than strictly causal. One does not have to drink the mystical Kool-Aid to entertain such a notion. When the quantum physicist talks about how two quantum-entangled particles, separated by a galaxy, can still be in instant sync with each other, we don't make accusations of magical thinking, and yet that is exactly what synchronicity is.

In fact, one can argue that science itself has been progressively destroying the deterministic, monist worldview. It began with Isaac Newton, who accidentally proved that the world was *not* a machine. His laws of gravitation were manifestly true, for they explained observed phenomena with unerring accuracy. But there was a problem: his laws relied on a mystical force called gravity. As Noam Chomsky explains, gravity "does not involve contact. So I can . . . move the moon by lifting my arm . . . [C]ommon sense tells us that I can only make something move by touching it . . . But Newtonian physics said, no, that's not true. There's an occult force that allows you to make things move."

Many of Newton's contemporaries were hypersensitive to any effort to introduce the supernatural back into science, and thus they loathed his theory and its appeal to something that was as seemingly magical as gravity and would say so with biting invective. Newton, a devout Christian, defended himself by maintaining that there was a purely mechanical force under there somewhere; he just didn't know what it was. Yet Chomsky's point about how you can move the moon by raising your arm is as true as it is prosaic. As he sees it, Newton's laws destroyed dualism, not by disproving the immaterial mind, but by showing that the commonsense mechanistic world doesn't exist at all. Even today, we don't know how gravity works; we can only describe its effects. Magnetism was another challenge to the "universe is a clockwork" worldview, another mystical force that is still beyond our understanding. But it didn't stop there: the more science learned, the more nothing was as it seemed. We discovered that what we see and hear aren't really there as we perceive them; rather, they are just mental constructs formed by the firing of neurons in our brains. We don't really "see" anything, and thus we have no idea how the world really looks and sounds. Later we busted open the atom, which we had thought to be the fundamental building block of the universe, only to find that it was full of all kinds of smaller things beyond our understanding, so fantastically strange that we christened them with Dr. Seussian names like quarks and squarks. When we went deeper into those, we found that there are no solids, only waves. This seemed odd because the world sure feels solid. It feels dense as well, but that also turns out to be merely an illusion, for science tells us that if we could remove all the empty space from the billions of humans alive today, we could all fit comfortably in the space of a sugar cube. Physicists claim that the universe is largely comprised of some mysterious substance called dark matter, which we cannot see, only infer, and that the expanding of the universe is powered by something called dark energy, which we also know only through inference and cannot explain. The list of the ways that science has disproven the mechanistic view of the world goes on almost without end: there are many dimensions, not just three or four; a butterfly in Brazil can cause a blizzard in Bethesda; the cat in the box is both dead and alive; observing reality changes it; if you travel fast, time slows down; and finally, matter exists but so does antimatter.

Synchronicity is about things that influence other things even though they are not causally connected. It sounds like mysticism, but that's the point—so does science. If the two particles in quantum entanglement act in accord despite being separated by interstellar space, why couldn't sticking a needle in a voodoo doll cause a distant person pain? If a lodestone, a naturally occurring magnet, can attract iron, why couldn't crystals heal? If the moon can raise and lower the earth's oceans—a fact we've known for four thousand years—why couldn't stars affect us too?

Finally, the fourth belief is that the future takes place the way it does because of *free will*. Fate is about the future being written. Determinism is about it being an inevitable outcome. Synchronicity is about it being inscrutable and unknowable. But maybe none of those is the whole story, and the real thing that determines the future is free will.

Free will is a rabbit hole too deep for us to venture down very far. To some, free will is simply disguised determinism. Your brain is a machine, and its output—your choices—is never in doubt, as it is the inevitable outcome of its physical system. But it sure doesn't feel that way, does it? It feels like we have free will. When Samuel Johnson was asked if humans have free will, he replied that all theory holds that we do not and all experience holds that we do.

Free will could be the choices made by the emergent entity that arises from the superorganism of you. If this is the case, then is that determinism? Or synchronicity? The answer to that depends on whether the kind of emergence that is happening is weak or strong. In weak emergence, the abilities and behavior of the system may not be present in any of its members, but the behavior is a predictable outcome. In strong emergence, which many people believe doesn't exist, the behavior or ability cannot be so derived. An example of this is consciousness. There is nothing, it is argued, that can account for how a hunk of matter can have an experience of the world.

If free will is the driving force behind how the future unfolds, what does that say of our ability to master it? That depends on the degree to which we can predict, in aggregate, what people will choose to do. In the book *Seeing into the Future,* Martin van Creveld noted that "Thucydides was not primarily interested in forecasting what might take place. Yet he explicitly based his work, and his claim to eternal fame, on the assumption that human nature was immutable and would always cause what had happened in the past to recur."

So those are our four main ideas of why the future unfolds a certain way and not another: fate, determinism, synchronicity, and free will. And within us, they all live side by side, in a strange kind of synthesis. I will tell an embarrassing story about myself to illustrate: I consider myself a person of science, making my decisions in a rational fashion. Most of the time. Like Michael Scott in *The Office,* "I'm not superstitious, but I am a little stitious." I have made multiple trips to North Korea, and I was contemplating another one. I was in a restaurant, alone, eating Chinese food, trying to decide whether or not to go. An old, old woman, at the next table, pointed her craggy finger at the woman she was dining with—but really at me because I was right over her companion's shoulder—and said in a witchlike voice, quite loudly, "Dooooon't goooo." I didn't go.

Likewise, surveys of Icelanders reveal that a majority, often a very large majority, believe that elves really exist on their island. I've always been curious about that, and on a trip to Reykjavík I made a point of asking different people I met if they did in fact believe that. The typical response I would get was a sheepish chuckle, a shake of the head "no" and then something like this: "Having said that . . . I did have one thing happen once that was very strange," and then they would relate some inexplicable tale of an encounter that sure sounded elfin. When asked if they agreed with how their country routes roads around places thought to be inhabited by elves, they would usually say something Nordicly practical such as, "Better safe than sorry." Do they really believe in elves? Let me put it this way: Icelanders might tell you that only chance governs the roll of the dice, but they still blow on them for luck. Don't we all?

The great Enlightenment thinkers were people with a foot in two worlds, one old and one new. They were scientists in the modern sense, but they also blew on the dice. Newton invented calculus, but he really wanted to master alchemy. Science was replacing superstition as the way to explain the world, but there was bound to be some overlap between the two worldviews.

Each of these methods for determining the future is thousands of years old. There hadn't been a lot of innovation in the field. That is, until the Enlightenment happened and we added a fifth one, a completely new, purely mathematical method of divination based on—of all things—randomness, and upon which our modern world is now built.

The Problem of Points

As math problems go, it is ridiculously simple. Two guys, Harry and Tommy, are playing a game of chance for a sum of money. In this game, a coin is flipped, and every time heads comes up, Harry gets a point. For each tails, Tommy scores. They agree to flip the coin five times, with the first person to three winning the whole pot. However, their game is interrupted and cannot be completed (I don't know, maybe someone stole their only coin) when Harry has two points and Tommy has one. What is the fair way to decide how to split the pot?

Think carefully about this problem. We will be coming back to it repeatedly. But before you read on, think about how *you* would solve it. How would you split the pot?

You can solve this problem a few ways. First, you can say that the two final tosses would have gone one of four ways: HH, HT, TH, TT. Given that in three of those four Harry would win, he should get three-fourths of the pot. Or, you could say that for Tommy to win the pot, toss four has to be tails, which happens half the time, *and* toss five has to be tails as well, which will happen in only half of *those* times. Thus you can determine that Tommy has only a 25 percent chance of taking the pot and therefore should get a quarter of it. Same answer.

Today, it is the sort of problem that wouldn't even earn a middle schooler a smiley-face stamp on their homework. And yet, it stumped the greatest mathematicians of their day for centuries. Let me say that again very plainly: this is a math problem no one could solve until 1654 when two intellectual giants swapping letters in France finally managed to crack it through effort, debate, and introspection.

The two men were Blaise Pascal, a thirty-one-year-old French mathematician who had fame as a prodigy in his teens, and mathematician Pierre de Fermat, age forty-six. They were both fascinating characters: Pascal mastered a wide range of scientific topics at an early age and had even helped support his family by designing and selling a mechanical calculator when he was just eighteen. Fermat spoke six languages, wrote poetry, and was only a hobbyist mathematician—earning his living as a lawyer—and yet contributed more to the field than Pascal, who would abandon math in favor of theology and die before his fortieth birthday. The problem of points had been put to them by another Frenchman, Chevalier de Méré, and Pascal and Fermat began working on it.

It was a well-known problem, and several potential solutions had been offered, but they were all wrong. Before we get to the story of how it was ultimately solved, let's trace the history of the problem and the various incorrect solutions.

In 1494, a friar named Luca Pacioli, a friend of Leonardo da Vinci, posed it in a book called *Everything about Arithmetic, Geometry, and Proportions*. It was a fascinating book, giving the Western world double-entry bookkeeping; teaching basic algebra; including a multiplication table up to sixty by sixty, quite useful in a world shifting from Roman to Arabic numbers; and, perhaps least significantly of all, posing the problem of points. He didn't create the problem—it was already in widespread circulation by then—and didn't solve it. But forty years after the book was published, the Italian polymath Gerolamo Cardano attempted the problem and failed. Two decades after this, Italian mathematician Niccolò Tartaglia gave it a try but to no avail. Four decades later, another famous mathematician, Lorenzo Forestani, sharpened his pencil and gave it a go but couldn't crack it. Decades after this, a writer named Chevalier de Méré sent it to Pascal, who took up the question in a series of letters with Fermat. Together, they finally solved it in 1654, and in the process invented the field of probability itself.

That's the history of the problem insofar as we know. What of the incorrect solutions?

The first one, and by far the most common one, is that there is no way to solve it. That is, to maintain that it is not a properly formed math problem. "What's two plus two?" is a math problem, while "What will happen tomorrow?" isn't one at all, and this sure looked like that.

But even people who rejected that logic still embraced the idea that the outcome was *unknowable*. And, of course, they are right. So the next solution offered is to split the pot fifty-fifty. Since no one knows who will win, the fair way is to give everyone their money back. This isn't an appealing solution to whoever happens to be winning at the time, but this was a commonly proposed answer.

About the only thing that early observers got right was that if the score was tied, the funds should be split evenly. That doesn't really require any math. But another early proposed partial solution was that if one player had no points and the other player had any at all, the one with the points should get the whole pot. But this is obviously wrong. If the game is to be played to one million points and one player is winning 1–0, they shouldn't get everything.

But from there, the solutions get a little more reasonable. Pacioli, whom we earlier thanked for the first published account of the problem, allocated the pot based on the number of points each player had, without regard for how many they needed. This is also clearly the wrong solution. There should be a big difference between how the pot is divided if the score is 2–1 in a game to three versus if the score is 2–1 in a game to one million. Tartaglia's solution was a step in the right direction but also still obviously wrong. His idea involved comparing the size of the lead of a player with the total number of points they are playing to and working from there. But again, obvious problems emerge. A player up by three points at the very end of a game is in a better position than a player up by three at the very beginning, yet

Tartaglia would split the pot the same way in both cases. He, too, concluded that perhaps the problem was unsolvable, that there was no way to divide the pot that both players would regard as fair.

Cardano offered another solution that got a little closer to the truth: perhaps what matters is not the number of points each player has, but how many each *needs*, and thus the winnings should be allocated in proportion to that. The premise is correct, but the solution is not. Say you are playing coin tosses to ten heads or ten tails and the score is nine heads to one tail. If the game is stopped while one player needs just one point to win and the other nine, do you give the player who is ahead 90 percent of the pot? No. The only way for the player with one point to win is to throw nine tails in a row, which happens only one time in 512 attempts. Thus, the real allocation in that case is 511/512 of the pot for the player with nine points and 1/512 of the pot for the other player.

Why did it take so long to solve such a simple problem? It isn't as though math waited until 1654 to be invented. Algebra was three thousand years old by then, and the Pythagorean theorem two thousand years old. Centuries before 1654, Chinese astronomer Guo Shoujing calculated the length of a year to within twenty-five *seconds*, and Persian mathematician Jamshīd Al-Kāshī calculated pi to seventeen digits. But if you had asked Guo, Al-Kāshī, or any mathematician who had ever lived up to that time, to solve the problem of points, you would have just gotten shrugs.

The main hurdle in solving the problem of points was not mathematical but conceptual. The solution required a new way to think about the future and how it would—or, more importantly, *could*—unfold. As Keith Devlin wrote in his book *The Unfinished Game*, when reading the correspondence between Pascal and Fermat, you "see the enormous difficulty these two great mathematicians had in grasping the very *idea* of predicting the likelihood of future events, let alone *how* to do so, you realize that what we nowadays take for granted was a huge advance in human thinking that came only through significant intellectual effort."

This book is interested in how we learned to see into the future. This section—"Dice"—is about how we developed a new mental model that involves considering multiple possible futures and the likelihood of each coming to pass. This is something entirely new, and the problem of points is, to us, a trivially simple embodiment of it, but to our forebears just a few centuries ago this was heady stuff. We will be returning to the problem of points and the 1600s repeatedly because that was the inflection point when our view of the future shifted. Our existing models for prediction were insufficient, and we knew it, as evidenced by the fact that the problem of points was both widely known and also unsolved. As Ian Hacking, a giant in the field of the philosophy of science, writes in *The Emergence of Probability*, "What is notable is not that problems on chance occur in early works of arithmetic . . . but that these books were quite unable to solve the problems. No one could solve them until about 1660, and then everyone could."

Making the conceptual leap of how to think about the problem of points was a huge step forward, but alone it is not a science of probability. Probability is not just the *idea* of how to solve the problem, but also the math to do it. Pascal and Fermat handled this as well.

Their creation of the mathematics went through an evolution throughout their correspondence. Pascal's first solution was what mathematicians call the method of recursion. He expressed it mathematically with proper notation, but for our purposes English will suffice. He said, in effect, "Let's assume they played just one more round in which they have an equal chance of winning. Does the problem get easier? If not, take each of those two tracks and see what would happen if they played just one more round." For instance, in our example, the next toss could be heads or tails. If it is heads, then Harry wins, and since there's a 50 percent chance of that happening, Harry should get to bank half the pot just for this alone. But what if it is tails? Then would another toss decide anything? If so, then keep allocating portions of the pot in the same manner, advancing the game one more hypothetical turn down each of the paths.

Fermat's method was a bit different. It is called the method of enumeration, and again, we can eschew the mathematical notation and simply say that enumeration is listing out all possible outcomes of the game. In our simple example, there are four ways the final two rolls could go: HH, HT,

TH, TT. In three of them heads wins, so heads deserves three-quarters of the pot. However, there is an interesting wrinkle that temporarily tripped up Pascal. He argued that HH and HT should count as just one possibility. Why? Because if the fourth roll is heads, then the game stops, Harry wins, and there is no fifth roll. So from this vantage point, shouldn't we just count three possibilities: H, TH, TT? Thus, the pot should be split two-thirds to one-third. The logic is tempting, and one can easily forgive Pascal's difficulties with it. You can do the problem that way, by saying there are only three possible outcomes, but H will occur 50 percent of the time while TH and TT will each occur only 25 percent of the time. Since only TT results in Tommy's victory, he gets just 25 percent of the pot.

Next, Pascal advanced the ball by beefing up the method and introducing something known as the arithmetic triangle, an easier way to solve somewhat more complicated cases. But both men realized that while these methods were useful ways to think about the problem, they quickly broke down in more complex cases. Imagine if it were three people flipping a three-sided coin and playing to a thousand points and the score was 231–445–312. How do you split the pot there? Well, the same logic applies, but it must be solved with formulas, and that's what they developed. Fermat introduced the negative binomial method for working the kind of problems where there are more than two players.

With their conceptual solutions backed up by all the relevant mathematical formulas, Pascal and Fermat had lit the fuse. Based on their work, other voices started offering new ways to calculate the solution to the problem of points. Columbia University's Prakash Gorroochurn laid out the histories of thirteen correct solutions to the problem of points, describing it as "not only the first major problem of probability but . . . also the one responsible for its foundation."

It is not impossible to find isolated problems that vaguely resemble the problem of points that were solved, or at least the correct answer was intuited, prior to Pascal and Fermat. But what rightly earned Pascal and Fermat their lofty place in history was that they invented an entire field of mathematics to solve this problem and a million more like it.

This was a time of scientific breakthroughs so foundational that they changed everything, quite different from our age of steady incrementalism. In 1543, Nicolaus Copernicus's *On the Revolutions of the Heavenly Spheres* was published. The book was so worldview altering that our casual usage of the word "revolution"—in its scientific, social, and political senses—was born from its title. In the blink of an eye, the Copernican Revolution changed the mental picture people had of the universe. Newton's laws, along with his calculus, would have a similar effect. The work of Pascal and Fermat was highly impactful as well, in large part because they were famous enough that when word of what they had created came out, the intellectual world was electrified. The two Frenchmen would contribute little more to the topic of probability, and both would be dead within a decade of their correspondence. But they had done enough, because thanks to them, the world went wild for probability.

What Did We Know Prior to 1654?

Was knowledge of probability really nonexistent when Pascal and Fermat were corresponding? For the most part, yes. The first inklings of understanding of the topic didn't appear until the mid-1500s, and then only in rudimentary form. Ian Hacking, when describing the 1865 book *A History of the Mathematical Theory of Probability from the Time of Pascal to That of Laplace*, wrote: "Its title is exactly right. There was hardly any history to record before Pascal, while after Laplace probability was so well understood that a page-by-page account of published work on the subject became almost impossible. Just six of the 618 pages of text in [it] discuss Pascal's predecessors."

The late development of probability might seem surprising given that math in general was pretty advanced. When you list out all the math knowledge we had at the time, it sure looks like the problem of points, as well as the science of probability in general, would be a cakewalk, but obviously it wasn't. This is analogous to a society knowing how to make the internal combustion motor and wheeled carriages, but never realizing it could build a car. Undoubtedly we have our own glaring blind spots today, which future generations will look at and wonder how we could have been so dense, such as how we have had luggage for centuries, but we put a man on the moon before anyone got the idea of attaching wheels to a suitcase.

However, I doubt probability as a science actually could have emerged much earlier. Sure, by 1654, all the parts were there to develop probability as a discipline, able to predict the future by assigning mathematical likelihoods to a range of possible outcomes. But we hadn't known it all for very long. So maybe it isn't so surprising that it wasn't until then that the conceptual leap on how to think about those problems happened to occur to someone with the mathematical chops to generalize that into a formal system. Still, it's hard to look at the following list of the four big things that we did know by 1654 and not wonder why no one had connected the various dots.

First, we knew a bit about probability with regards to games of chance. Much of this knowledge we owe to a fascinating guy named Gerolamo Cardano. He lived in the 1500s, and throughout his long life, he became a mathematician, philosopher, author, doctor, chemist, biologist, astronomer, astrologer, chess master, and gambler. There was evidently no better time to be a Renaissance man than the Renaissance. He also published over two hundred scientific works, including one on gambling called *The Book on Games of Chance*, which covered a bit of what we would call probability. He knew, for instance, that the odds of getting a six in a single roll of a die were one in six, obviously, but he also figured out that to compute the odds of throwing two sixes with two dice, you multiplied the odds to come up with one in thirty-six. So if anyone deserves a precursor award in probability, we should hand that ribbon to Cardano, although his pieces on probability weren't all accurate. Incidentally, the book also had a good deal of information on how to cheat at gambling, which he included—he claimed—solely to keep his readers from being swindled. Sure, Gerolamo, sure.

He is often shortchanged in the history of probability because his gambling book languished unpublished for over a century, until *after* Pascal and Fermat became pen pals. He did frequently lecture, so it is assumed he shopped his ideas around. But we mustn't overstate Cardano's contribution either. For one thing, he was writing about gambling, and specifically the odds in certain games. Everything he knew, you could get to with simple multiplication, unlike the alphabet-soup formulas required by real calculations in probability—the ones that Pascal and Fermat developed to solve generalized problems.

But consider this: Cardano was a genius. He is the guy who more or less invented the combination lock as well as an essential part of the automatic transmission in your car. He thought up the idea of using letters as unknown numbers in algebra, a topic he wrote an entire book about. He worked on solutions to cubic and quadratic equations, as well as square roots of negative numbers. Yet he tried and failed to solve the problem of points.

Second, in 1654, we knew a good deal about dice probabilities. Dice are ancient, vastly older than coin tosses. They are so ancient that the reason that they have pips instead of numbers is that when they were invented, there were not yet any symbols for numbers. Visitors to Pompeii today can see vivid frescoes of men playing dice. The fact that archaeologists keep uncovering loaded—that is, dishonest—dice from across antiquity actually suggests a deeper knowledge of probability than fair dice would. It wasn't just loaded dice either. We have examples of ancient dice where one value is used twice while another number is omitted. Sneaky. Don't try palming one of those into the next craps game you play in Las Vegas.

By Pascal's time, activities involving the rolling of three dice were commonplace. With three dice, you can roll any value between three and eighteen. That's obvious. But what also had been widely known for over six hundred years was that there were exactly fifty-six combinations that could come up: 1-1-1, 1-1-2, and so forth. There is a fortune-telling medieval poem whose fifty-six verses tell you your future based on the outcome of a single roll of three dice. We have evidence that by the 1200s it was also known that there were 216 unique permutations with three dice. The difference between the fifty-six combinations and the 216 permutations is important to understand. The former disregards the order of the dice. It counts the roll of a one, a one, and a two as a single entry, but the latter counts it as three, depending on the order, that is, 1-1-2, 1-2-1, 2-1-1. We'll see why that matters shortly.

By the 1400s, there circulated a belief among the gambling cognoscenti that tens were more often thrown than nines. This was puzzling, because there were the same number of ways to throw each of those rolls, that is, six ways. For instance, for ten, you could throw 1-3-6, 2-2-6, 2-3-5, 1-4-5, 2-4-4, and 3-3-4.

That's where the matter stood when the most unlikely of people enters our story: Galileo Galilei. In his role as First and Extraordinary Mathematician of the University of Pisa—which must have looked pretty good on his résumé—he was pulled away from staring at sunspots through a telescope by his patron Cosimo II de' Medici, grand duke of Tuscany. Cosimo, who was likely looking for an edge at the gambling table, asked Galileo to figure out if there was anything to this whole "tens are more common than nines" thing. Galileo no doubt thought there were better uses of his time. He dutifully obeyed, but he wrote it in the vernacular Italian, not wanting to sully scholarly Latin with such a pedestrian topic. He explained in an orderly and concise way why tens were more common than nines. While they both had the same number of combinations, six, they had different permutations. Ten has twenty-seven while nine has only twenty-five. This is in large part because one of the permutations for nine is 3-3-3, which can be rolled only one way.

But here is the puzzling part of the situation. It is unlikely that anyone would notice the difference in the relative frequency of tens and nines. To get 95 percent confidence that tens were in fact more common than nines, you would need to roll three dice five thousand times, carefully tallying which of each 216 permutations you get. So how did people know this? This isn't the only such case where gambler's lore is seemingly beyond mathematical knowledge. Poker players ranked flushes higher than straights before it could be proven that the latter were more common. In *Games, Gods and Gambling*, F. N. David expresses a belief that Galileo's patron "had gambled often enough to be able to detect a difference in probabilities of 1/108." That might be true, and if so, it was quite a feat. Others maintain that the gamblers knew the math but kept their mouths shut, and this could also be, but if it were the case, why would the *edge* this gave the gamblers become common knowledge but the *technique* for discerning it remain unknown for centuries? In a paper charmingly titled "The Role of Roguery in the History of Probability," David Bellhouse studied this question and concluded that "contrary to the accepted folk-lore . . . gambling itself provided very little stimulus to the development of probability theory. In the other direction, the development of the probability calculus had a profound effect on gambling, namely in the formation of a strategy of play."

You may be thinking that all of this is pretty indistinguishable from probability as a science. But recall the probability question I posed earlier: imagine the problem of points if it were three people flipping a three-sided coin and playing to a thousand points and the score was 231–445–312. How do you split the pot there? All the techniques employed by Cardano, Galileo, and the myriad unnamed gamblers would never have been able to answer that question. All they were doing was enumerating combinations of simple things like dice and cards, then counting them. All Galileo did was to count permutations and note that there were more ways to roll a ten than a nine. It wasn't formula based, nor did he develop it into a generalizable science. But Cosimo hadn't asked him to do that, just to solve a problem. And Galileo did more than he was asked, and undoubtedly could have done more still, but the moons of Jupiter were not going to count themselves.

Third, in 1654 we had sophisticated business transactions occurring around the world that required a basic understanding of risk assessment. There were stocks, futures, options, lotteries, and insurance policies, all of which are obviously risk centered. Less obvious examples are even more abundant: the famous tulip mania that occurred just before Pascal and Fermat's correspondence can best be thought of as a market where people bought and sold risk, not tulips.

Insurance, which today is a multitrillion-dollar industry, is a good place to focus for our purposes, because it requires a quantification of the risk of certain future events and thus is not totally dissimilar to the problem of points.

In the ancient world, merchants frequently had to borrow money to finance their trade, putting up the cargo as collateral. In the event the ship was lost at sea, the loan didn't have to be repaid. The practice is called bottomry and is dealt with extensively in the four-thousand-year-old Code of Hammurabi. The interest rates the merchants were charged were both high and highly variable, depending on the risk of the specific voyage. Thus, the loans could be interpreted as a form of insurance, and the excess interest over the prevailing rate of the day was the insurance premium. But it wasn't really an insurance industry per se, since, as economic historian Paul Millett of the University of Cambridge points out, in ancient Greece and elsewhere, "traders always sailed with their cargo. So if the ship went down, there was always a good chance that they would go down with it, making the question of insurance irrelevant." However, by the end of the thirteenth century, Italian merchants had stopped traveling with their cargo as they became more sedentary "desk jockeys" than adventurous traders. In that environment, a bona fide insurance industry emerged in the mid-1300s.

Once cargo was being insured, it was logical to insure other things. An entire industry emerged that was designed to bear financial risk for a price. Perhaps there was even an intuitive sense that the more risk that is pooled, the less risky it becomes, but we don't have evidence of anything like that explicitly acknowledged. The range of what you could insure back then rivals our options today. Three mundane examples of the thousands we know of: a traveler bought kidnapping insurance for a trip to Sardinia that would pay his ransom for about 1.5 percent of the cost of the maximum payout; a doctor bought a policy on his pregnant slave not dying in childbirth with a premium of 8 percent of the payout; and a creditor bought life insurance on his debtor, an archbishop, for a premium of 10 percent of the insured amount.

Insurance can sometimes be indistinguishable from gambling. When a ship owner buys a policy on the safe arrival of his cargo, that's insurance. But what quickly emerged were situations where neither the buyer nor the seller of the insurance had any economic interest in the cargo beyond the insurance contract. Person A might sell B an insurance policy for the safe arrival of a certain ship into harbor merely as a financial sporting matter. In a 1912 book on the history of life insurance, the University of Manchester's A. Fingland Jack wrote that "the custom grew up of making wagers, under the guise of insurance, on the lives of others, where no bona fide interest could possibly be pretended to exist. Wagers, for example, against the death of the king or the pope within a given period could only . . . be the outcome of the gambling passion." Incidentally, today you have to show that you would be financially harmed by someone's death to own a life policy on them, lest everyone goes and buys insurance on their crazy daredevil uncle who they have good reason to suspect will meet an early end.

Does the development of insurance imply an understanding of the fundamentals of probability? Not really. As James Franklin writes in *The Science of Conjecture*, "While the writing of insurance contracts obviously implies an explicit quantification of risks, it is one thing to do this and another to be aware of doing it. There is no theoretical writing on the subject by the merchants, and the conceptualization of risk is not usually apparent in the documentary evidence." The premium charged for the policy is obviously based on a probability—a specific probability, in fact. But that does not a science make. However, it does lay the groundwork for thinking of the future probabilistically.

The fourth thing we knew in 1654 was related to variable gambling payouts, the way a certain horse might pay twenty to one in a race. In a paper called "The Language of Chance," Bellhouse and James Franklin document the occurrences of odds in English literature all the way back to Chaucer's day. References to this idea appear in several of Shakespeare's plays, decades before the Pascal/Fermat letters. The Bard used the phrases "four to one" and "five to one" once each, "ten to one" six times, and "twenty to one" four times, all the same way we do today, such as in *Much Ado About Nothing* when during a discussion as to whether the prince can be legally detained, a character says, "I'd bet five to one that he can—ask any man who knows the acts of Parliament." The paper concludes that "the use of odds to describe probabilities explicitly, in terms of numerical odds, and implicitly, in terms of the balance of probability, appear to have been widespread in the culture. Consequently, it is difficult to imagine that no one in the culture had actually made an elementary probability calculation in a simple game of chance."

But no one took all this and made it into a science. It seems no one said, "The future can unfold in various ways with various likelihoods, then those things can unfold these different ways," and—here's the important part—turned that into a mathematical system. The knowledge that one ship in ten might sink, thus insurance should cost at least 10 percent of the cargo value, is something a merchant could explicitly or implicitly know—but still not know about probability as a science.

Why Not Earlier?

Looking at the list presented in the previous chapter, it's not just fair to ask why the science of probability hadn't been invented before 1654, it's almost impossible not to. A few reasons:

First, probability isn't actually all that intuitive. It is to us, because it is all sorted out now, and we can just learn the techniques, but imagine trying to invent them. For instance, consider this simple problem. What are the odds of rolling a six in a single roll of a die? One in six, right? But what if I give you two rolls? What are the odds of rolling a six in at least one of them? The obvious, intuitive, and incorrect answer is that since you have a one-in-six chance on the first roll and a one-in-six chance on the second roll, then you must have a two-in-six chance overall. Ninety-nine people out of a hundred would probably give you that answer until you ask a follow-up, "Well, what about in six rolls then? Is it therefore a six-in-six chance? Or how about in seven rolls?" To solve, you have to figure out the odds of not rolling a six and then multiply them and subtract that value from one. Who would have guessed? No one, it turns out, until recently. But once one person did, it became fair game on the SAT. Or there's the old problem of, "What are the odds that in a room of twenty-three people, two will share a birthday?" The answer is that they are better than fifty-fifty. It doesn't seem true, but when you explain that person A has twenty-two chances to match someone, and person B has twenty-one additional chances to match someone, and person C has twenty more chances, it suddenly makes sense.

Second, there just wasn't much of what we now call data a few centuries ago. Today's world is full of data collection. We obsessively count and log everything. We have trillions of sensors—that's millions of millions—each of which deals with numerical data. But what was floating around in 1600? What data collection was being performed? The closest things might be parish records of births and marriages. True, back then people counted money, livestock, ships, and, well, other people, but counting isn't even really math, and it sure doesn't get you closer to probability. Outside of gaming, where would you apply probability in, say, the year 1200?

Third, Roman numerals were still being widely used right up to Pascal's time. Heck, they're still used today. Just ask Queen Elizabeth II next time you bump into her at a *Rocky III* screening or at Super Bowl LV.* But imagine if that were your only system. What is DCIV divided by LII?

While Arabic numbers began to replace Roman ones in Western Europe around 1300, the process was agonizingly slow and didn't really become ubiquitous until about 1600. Why is this? For one thing, it's hard to get people to change practices as basic as how they count. During the French Revolution, decimal time was mandated. A day was ten hours, an hour one hundred minutes, and a minute one hundred decimal seconds. To say it did not catch on would be a gross understatement. Imagine trying to get that passed through Congress today. In addition, there were other barriers to the adoption of Arabic numbers. For example, Peter Bernstein writes in *Against the Gods* that "Florence issued an edict in 1229 that forbade bankers from using the 'infidel' symbols. As a result, many people who wanted to learn the new system had to disguise themselves as Moslems in order to do so."

Sophisticated math was impossible using Roman numerals. While they had a limited way to deal with fractions, nothing could match the Arabic system. Plus there is the whole idea of zero as a place value in numbers,

---

* Super Bowl 50 was a onetime exception, since Super Bowl L doesn't really have the same pizzazz.

which the Roman numeral system also lacked. The zero in the number 101 makes all the difference in the world because without it you just have eleven. Or as Jethro put it in an episode of *The Beverly Hillbillies*, "Funny thing about them naughts: by themselves, they ain't worth shucks, but you lay 'em up behind a number, and them rascals is dynamite."

But even Arabic numbers themselves aren't much use without notation, and that came even later. When Pascal and Fermat exchanged their letters, the equal sign, the multiplication sign, and the decimal point were less than a century old. The radical symbol ($\sqrt{}$) for square root wasn't even old enough to order a beer, nor was the percent sign (%). As James Franklin wrote, our basic tools of math such as "decimal notation, negative numbers, algebraic symbolism, graphs on the Cartesian plane, and logarithms were only tentatively known here and there before 1600 but were common property by 1650. Without them, it is a struggle to do mathematics using essentially only natural language and geometrical diagrams; with them, everything becomes possible."

Fourth, the easiest way to study the basics of probability centuries ago would have been with dice. But dice were also used extensively in divination, which might have made them seem inappropriate for scientific inquiry. This analogy is not perfect, but it could have been akin to a scientist today using a Ouija board or Tarot deck as part of a real scientific inquiry. Even if the experiment was purely mathematical, the optics are problematic. In addition, dice were used in gaming and gambling, not exactly what are thought of as fertile fields for mathematical progress.

Fifth, it has been argued that the rise of the modern economy from, say, the 1400s onward required more sophistication in mathematics, and that this is what drove advances in the field. Sure, math as a pure science was interesting, as was math in support of scientific inquiry such as astronomy. But to really make it take off, it had to become profitable.

All of these reasons are valid, but even taken all together, they don't really explain how we could have known so much and yet still weren't able to connect the dots to understand our modern conception of probability. All the math knowledge we needed was already there. It was the conceptual framework that was lacking. There were five things we didn't yet understand, five things more fundamental than all the reasons we just went through. They weren't ideas that we can pin to a place and point in time but five foundational truths birthed in the Enlightenment, which gradually developed over time. Even today, we do not have mastery of them, and they are still matters of hot debate. But at the height of the Renaissance, when Pascal and Fermat were trading letters, we began to see them in outline, and they are what allowed us to make the intellectual leaps to predict the future in a new and exciting way.

The next five chapters will explore each of these ideas and its implications to us today. All five are scientific truths, but I am struck even now by how much they sound like magical thinking—though no more so than gravity. I'm going to lean into the mystical language just a bit because it best captures the wonder of how we learned to do something that still seems magical: to see into the future.

A Numerical Reality

The first of the five ideas we discovered that allowed us to see into the future is that we inhabit a numerical reality. Numbers hold huge sway over our daily lives, and they are our main tool for seeing the future. This is pretty interesting since we are not by nature particularly numeric, unlike, say, computers. Computers *only* know how to do math, and everything they do is at its core numeric. Our existence is a sticky, tactile, biological one. To us, numbers are a foreign language that we had to master to understand our world and predict the future.

Consider the number of numbers that have influence in your life. We can start with the basics: your age, birth date, height, weight, BMI, income, credit score, bank balance, and the number of email messages in your inbox—to name just a few. Then there are all the numbers you come in contact with: the temperature, chance of rain, relative humidity, speed limits, flight times, and the rest. In your house, you have a thermostat setting; your microwave uses a certain number of watts, as do your light bulbs. Your house has a square footage, a construction year, an annual tax bill, and a mortgage with an APR. Cars deal in numbers, too: they have a model year, mileage, engine displacement, number of cylinders, fuel economy, and a psi setting for your tires. Schools are no different—they are ruled by class times, grades, test scores, class ranks, and credit hours. The list of numbers that affect our lives goes on indefinitely.

This is a new phenomenon. As J. F. Bierlein writes in *Parallel Myths*, "even today it is not uncommon for a person living in a traditional culture to not know his or her chronological age; it simply doesn't matter." Antiquity is bereft of quantifiable statistics. We don't know, for instance, the approval ratings of ancient leaders nor the mortality rates of their soldiers nor their average shoe size.

The *idea* of numbers had to be invented. The notion that five fingers have something in common with five rocks, and that this something is an abstract idea that could be called fiveness, is not at all obvious. More than one ancient language used the same word for "hand" as "five," perhaps suggesting an origin of thought about numbers. Today, we manipulate numbers with a casual ease that would have seemed alien to humans until recently. What changed? We discovered that the universe can be deciphered, but that its inner workings are not words but numbers, and their deepest secrets are ratios, that is, the relation of numbers to other numbers. Thus, to achieve mastery over our world, we had to learn a new way to perceive reality that is fundamentally numeric. We figured out that if we want to measure the pull of Jupiter on one of its moons—Titan, for instance—we can't do that in the vernacular. We have to convert Jupiter and Titan each to a number called mass, then convert the space between them to a number called distance. Only after having done that can we perform calculations using an entirely different set of symbols than we use in our daily lives and arrive at our answer.

Think about the sheer wonder of it all. Isaac Newton figured out something about planetary motion that can be expressed with five letters, four symbols, and three numbers: $F = (Gm_1m_2)/r^2$. That's it. And amazingly, it can account for the orbit of Neptune as well as—and here's the crazy part—seemingly unrelated things in our everyday life, from lawn darts to ocean tides. Science repeatedly delivers formulas that explain specific observed phenomena but are also found to be useful for myriad other purposes, allowing us to make highly accurate predictions, right down to being able to land a two-ton probe on a bubblegum wrapper on Mars if we so choose and if, of course, there happens to be one there.

This is not to say that people in antiquity didn't understand math. Ancient Egyptians calculated areas of land with great precision. The Greeks knew—and were troubled by—the fact that the square root of two, that is, the length of a diagonal line through a square, was irrational. The Romans were masterful surveyors and engineers, deftly manipulating numbers in those spheres. The Chinese came up with negative numbers; the Babylonians had a zero as a place-value holder; India gave us our number system; and a thousand years ago, algebra emerged in the Muslim world.

But that's different from the idea that reality itself is numeric. That idea is only a few centuries old. Galileo expressed it prosaically: "Mathematics is the language in which God has written the universe." You can describe gravity with words, but you cannot get very far without numbers and math. We didn't discover this truth about the nature of our universe earlier because although we had math, we didn't have enough to describe how the world operates. $F = (Gm_1m_2)/r^2$ isn't simply a formula; it is a statement about the nature of reality. It is more than algebra, more than geometry, more than arithmetic. It is a declaration, a discovery, a fact, and almost a sacred truth. That's what it means to say reality is numerical.

Why are physical laws inherently numeric? One explanation—which I personally don't believe—is that this is proof of the simulation hypothesis, that is, that we don't really exist at all but rather are just characters in a computer simulation. The fact that the universe seems to only speak the native language of machines is seen as proof that the universe, too, is such a machine.

There is an old joke about an economist who sees something working in the real world and wonders if it is possible for it to work in theory. We can chuckle at this because, well, it's a social science after all, and frankly we don't expect all that much precision from its predictions. But our scientific theories, well, we demand far more from them. We expect such a theory to be elegant and for it to bear out under empirical observation. And often it does, much to the surprise of its discoverers. *But why?* Even Albert Einstein found this question a real head-scratcher, remarking that "the most incomprehensible thing about the universe is that it is comprehensible."

In 1960, a physicist named Eugene Wigner published a paper asking these very questions, called "The Unreasonable Effectiveness of Mathematics in the Natural Sciences." He looked at the problem through a number of lenses but ultimately came to no conclusion, just a realization that "the miracle of the appropriateness of the language of mathematics for the formulation of the laws of physics is a wonderful gift which we neither understand nor deserve. We should be grateful for it and hope that it will remain valid in future research."

As our world became more numerical, we demanded ever more precision in our numbers. They were, as it turns out, the key to our mastery of seeing the future. Nowhere is this plainer than in timekeeping. We used to just have sundials and hourglasses. Was anyone ever five minutes late in that world? Who could call them on it, telling them they were seven hundred grains of sand late to work that day? Eventually we got clocks, often bearing admonishments on their dials not to waste time. They had just one hand to show the hour, which was precise enough for most purposes. Then we added the minute hand. Then, the second hand. The stopwatch function on your smartphone measures to the hundredths of a second, in case, one supposes, you are unexpectedly called upon to judge an Olympic sprinting trial. Now, we make clocks so precise that if there were a major earthquake in California, we could detect it by its gravitational distortion of time.

In the *Matrix* movies, the matrix is represented as green source code flying across computer monitors. Those in the matrix see a normal-looking reality, but their allies monitoring them from a ship can see the code and, to a degree, read it. Our reality looks a bit like that. What we have been able to do with the red pill of math is to see the code, the math, under our reality. We flip between these worlds all throughout our days, hardly even noticing.

Determinism's Limit

Our second big idea that helped pave the way to probability involved coming to understand the inherent limits of determinism. As philosophies go, determinism seems like the embodiment of common sense. What could be more obvious than the idea that everything has a cause, each of which in turn has a cause, all the way back? It really does make sense that if you rewound the universe to the beginning and pressed the play button, you would see the exact same movie again.

There is of course a standard critique of determinism that says that at the quantum level, there is true uncertainty—that there are events without causes. This is likely true, but not relevant to our purposes. We don't live our daily lives in a quantum reality, but a (mostly) Newtonian one. That's not the limit we're talking about here.

Determinism's promise is about its ability to predict the future. It posits that if you take accurate measurements, then apply the right formula, you, too, can see the future. As your measurements get ever more precise, so your predictions will be ever more accurate.

However, as commonsense as all that sounds, it's wrong. The fascinating truth is that for complex systems, reducing the error in your data isn't guaranteed to get you more accurate results. Consider the story of Edward Lorenz, a meteorologist in the 1960s who used a computer to model the weather. He would set initial conditions in the computer, and then it would print out the weather conditions as they hypothetically unfolded. What he accidentally discovered was that if he took a printout of the data from, say, halfway through the simulation, then entered that as the initial conditions to the model and reran the simulation, it didn't produce the same outcome. It wasn't even close. How could this be? The computer should produce the same output every step of the way for a given set of initial conditions. The answer was that his model was running using numbers to six decimal places, but the printout only printed to three. He didn't know this, but even if he had, what difference would it make whether he entered 0.347442 or 0.347? Turns out it is a great deal of difference. This was dubbed the butterfly effect, and Lorenz had discovered chaos theory.

What if Lorenz had used four digits or even five? Strangely, his results might have been even further off. In chaos there is no relationship between the degree of error in setting the initial conditions and the accuracy of the predictions. A giant eagle flapping its wings in South America doesn't cause a bigger tornado in Toledo than that of a butterfly. Luckily most things in life aren't like this. If you are a bit sloppy in measuring the ingredients of a cake you are baking, then the cake probably tastes fine, but less so as the error increases. If baking were chaotic, then a single grain of salt too many would make the cake poisonous, but two extra grains would make it perfect, and three would result in it tasting like fried chicken. While there is no actual randomness in chaos—it is completely deterministic—there is unpredictability if measurements are off by even the smallest degree.

This is even worse than it sounds. For certain cases, such as the weather, truly accurate measurement is literally impossible. Even if your numbers are accurate to a hundred digits, there are still digits 101 to infinity to throw your model way off. To use the butterfly metaphor again, if a butterfly really could cause a tornado, then to predict the weather you would need to know the location, weight, speed, and orientation of every butterfly on the planet. This is why long-term weather prediction may actually be impossible.

So what was determinism's limit? That for most complex systems, the degree of error that is present in measurements is enough to introduce substantial and unpredictable errors in prediction. Simple cases, such as plotting where a cannonball will hit, or even when to launch a space probe so it can slingshot off the gravity of another planet, can be done quite effectively using determinism since small errors are not able to compound on themselves the way they did for Lorenz. But for many areas in which we want to predict the future, we cannot start with a set of initial conditions and figure out what will happen, because setting initial conditions is impossible.

Enlightenment thinkers didn't know this, of course. Their hope was that laws could be found not just for predicting cannonballs in flight but for biology, sociology, economics, and the rest. Why shouldn't there be such laws, since they are natural phenomena as well? The answer is that while determinism is handy for relatively simple predictions, it breaks down with the kinds of complex systems that power the world. How in the world can we get around that conundrum? That is our third big idea: randomness's miracle.

Randomness's Miracle

Earlier we looked at different belief systems about why the future happens the way it does. One element lacking from all of them is randomness. Whether things are fated to happen, or happen through synchronicity or necessity, or are chosen through free will, no views had a notion of the future that was random. To be clear, people have long believed that certain things were *unknowable*, but that is quite a bit different from randomness. As we will soon see, randomness not only drives the future but is our best tool for predicting it.

What is randomness? There are no good consensus definitions, but broadly we can say that random events have no pattern and cannot be individually predicted. They are independent and not influenced by any prior events; otherwise, they could be predicted.

Depending on how you define randomness, it is either everywhere, uncommon, virtually nonexistent, or impossible. Let's look at each of those:

Everywhere. Practically speaking, randomness is everywhere, not just in dice rolls and coin tosses. It's in noise and static, sunspots and gamma rays. If you go to the beach and scoop up a pail of sand, there are perhaps a million grains. If you took the time to count them all, the last digit in that number is about as random as things come. If you were to take a piece of poster board and use a marker and a straight edge to divide it into ten equal parts, you could place it in your backyard, watch raindrops hit it, and write down which section of the board that each one landed on. This would be a fine random-number generator.

Surprisingly, randomness is also in social phenomena everywhere, a fact we are about to spend a fair amount of ink exploring. Even you are a random phenomenon. Something on the order of a trillion generations separates you from that first spark of life that appeared all those years ago. Had just one of those happened any differently, well, poof! No you. Randomness is perhaps the fundamental truth of the physical world, beginning with when the Big Bang scattered matter throughout the universe.

Uncommon. Actually, randomness is pretty uncommon. Squint long enough and you can find little patterns in things that are supposed to be random. Coin tosses slightly favor heads, dice rolls are measurably sensitive to initial conditions, and even our poster board in the field suffers from the fact that raindrops are naturally spaced in a way randomness shouldn't be. Lots of things look random but aren't. What we find in the real world is that many things cluster around a mean, so they aren't random at all. Human behavior is often full of all kinds of subtle patterns that preclude randomness. Ask people to pick a random number between one and ten, and a quarter of them pick seven.

Randomness may not even lurk in the infinite, seemingly unpatterned digits of pi. We've calculated pi to tens of trillions of decimal places. Wouldn't a list of every millionth digit be a random string of numbers? We don't really know. We can say that the first trillion digits look pretty evenly distributed, with only about a million more of the most common digit, 8, than the least common, zero. A million sounds like a lot, but it is only one one-millionth of a trillion, so it is tiny in this context. But pi probably *shouldn't* be random anyway. It is, after all, a very specific value representing a very real-world thing.

Randomness is so fickle that it is hard to fake. Theodore Hill of the Georgia Institute of Technology gave his math students the following homework assignment: either flip a coin two hundred times and write the results or pretend to flip a coin two hundred times and write down the hypothetical results. The next day, he effortlessly picked out the real flips from the fakes with high accuracy. The trick? A real coin toss will almost always have a run of at least six heads or tails. Fakers are seldom bold enough to insert that many repeated results in their "real-looking" numbers. Poor random-number generators are relatively easy for hackers to break. The Nazis were, fortunately, not very good at generating truly random keys for their secret code machine Enigma, opting instead for the same letter repeated multiple times or their own initials. In essence, they used the word "password" as their password, allowing the code to be broken and shaving two years off the war.

Generating random numbers is a tricky affair. Computers can't really do it since they are deterministic in nature. A program to create random numbers will generate the same numbers on a thousand computers unless they at least use something randomish from which to generate their results. Often, the movements of the mouse or the number of milliseconds between a user's double click on an icon can be roped into use.

Human activity throws off lots of numbers, but they aren't random, or so says Benford's law, which states that we produce more numbers that begin with ones than twos, more twos than threes, and so forth. Why? Think about the first twenty numbers, one to twenty. How many of those start with one? Eleven by my count, or just over half. But that percentage falls as you get close to one hundred. But then, bam! You hit one hundred, and you get a hundred numbers in a row that start with one, pushing the percentage back up to over half. Then you hit one thousand and get a thousand numbers in a row starting with one. So, sure, across infinity numbers, it all levels out, but our day-to-day lives aren't constructed of infinity numbers; they are bound sets, like our ages, bank balances, and so forth.

Virtually nonexistent. Maybe just a few things are truly random, and they occur only at the quantum level, in phenomena such as radioactive decay. The half-life of uranium-238 is about 4.5 billion years. That means that if you had a bucket of the stuff sitting around your living room, it would constantly be emitting alpha particles as it slowly changed into thorium-234. After 4.5 billion years, exactly half of uranium will have become thorium. Ahhh, but which half? There's no way to know. One can almost picture all those nuclei sitting around, side-eyeing each other, wondering who was going to decay next. But scientists tell us there is no way of knowing. Literally no way of knowing. This is perhaps the only real randomness in all the universe.

Impossible. Or maybe it is completely impossible. We just theorize that things at the quantum realm such as radioactive decay are purely probabilistic, not mechanistic. We don't actually know that. Although that has been the widespread belief for the past century, we can't really say this with any certainty. After all, Einstein had his reservations about the whole thing, which inspired his famous quote about God not playing dice with the universe, and Erwin Schrödinger, of cat fame, believed that there was an underlying deterministic reality to quantum mechanics.

In the end, it doesn't really matter all that much for almost all purposes. Consider coin tosses, which have been used as randomizers about as long as there have been coins. The ancient Romans called it *navia aut caput*, that is, ship or head, since a typical coin might have a ruler on one side and a warship on the other. However, the flip of the coin is not actually random. Imagine a coin, heads up, on a table. With your thumb and index finger, you grasp the rim of the coin, raise it a tenth of an inch, and let it go, performing the world's laziest coin toss. With 100 percent certainty, I can say it will come up heads. There is unquestionably no randomness there. But what if you raise it a foot and drop it? I did this, and it still comes up heads virtually all the time. But what if you throw it out the window of an airplane at thirty thousand feet? Well, now we don't know. We are back to butterflies and tornadoes. But here's the kicker: The coin dropped from six miles is no more random than the one dropped from a tenth of an inch. We just don't have the data nor the mental agility to suss out the latter case, as it is governed by chaos. Yet the probability of calling the coin correctly goes from 100 percent to fifty-fifty as the distance of the drop increases. From our perspective, and for all intents and purposes, it becomes random. Neither a computer nor a mathematician could look at the results of two hundred tosses from a plane versus two hundred random results from a quantum randomizer and tell which is which.

The question of true randomness is important only up to a point. The real question is whether future events—tosses, in this case—can be predicted with any more accuracy than truly random ones. If not, then whether it is "really" random is moot.

Why does any of this matter? It is because randomness is the basis for how we predict the future. Could anything be more counterintuitive? Randomness would seem to be the *least* useful thing for helping predict the future. It sounds like something from DC's Bizarro World. But it does help us immensely because—this is the "miracle" that is referred to in this chapter's title—there is an enormous amount of predictability in a group of random events. That, too, is a strange sentence to even type, let alone wrap one's head around. How can randomness be predictable? Let's explore that.

If you flip a coin a thousand times, how often will it come up heads? You probably don't even need to think about it at all—around five hundred, give or take. But how do you *know* that? You probably haven't actually flipped a thousand coins and counted or met anyone who has. While five hundred seems like the obvious answer to us, it is only because we were birthed in a world that thinks this way. A far more reasonable answer to the question seems to be, "There's no way to know." How could you? It seems a question beyond scientific inquiry—or even rational analysis. If you believed the coin tosses to be random, how could you say anything intelligent about how they will turn up? Cicero saw this thousands of years ago when he asked, rhetorically, "How can an event be . . . predicted which occurs . . . as a result of blind chance?" Wouldn't throwing five hundred heads be about as likely as throwing eight hundred heads or two hundred heads? Wouldn't it seem, strictly intuitively, like the number of heads should be all over the map?

Yet it isn't. If you flip the coin a thousand times, there is less than *one chance in a billion* that you get more than six hundred heads or tails. Waaaaay less than one in a billion. It simply isn't going to happen in a practical sense. How can something that certain actually be random? Before the 1650s, no one could have even posed that question, let alone answered it.

After the 1650s, it became widely noticed by scientists that random events like coin tosses and dice rolls did seem to follow long-term patterns with almost unerring accuracy. And not just dice and coins, but all kinds of other things that were seemingly built from randomness, such as the ratio of male to female births, which we can think of as biological coin tosses. The most common explanation at that time, offered without apology, was that such things show that the Almighty doesn't just govern the orbits of the planets or the passing of the seasons but even holds such regularities in check. The fact that the relative number of male births to female births was noted to be quite stable from year to year could more easily be seen as the hand of God than some strange property of randomness, especially since there were always more male births, a divine allowance, it would seem, for the fact that a surplus of men would always be needed because of inevitable losses in war. The seventeenth-century French mathematician Abraham de Moivre expressed such a sentiment, stating, "If we blind not ourselves with metaphysical dust we shall be led by a short and obvious way, to the acknowledgment of the great Maker and Governor of all."

Whatever the force behind it was, mathematicians in the 1600s got serious enough about understanding these long-term patterns. They rolled dice or flipped coins a thousand times and wrote down the results. What they learned was that while no individual roll or flip could be predicted, the total of the outcomes could be, with astonishing accuracy. And the more times the coin was flipped or the dice was rolled, the more predictable the overall outcome was. That was the miracle, that randomness could be predictable.

The realization that there was such stability in randomness is more than trivia because it applies to more than coin tosses and dice rolls. Once we realized that randomness was, for practical purposes, everywhere, we rightly inferred that predictability was everywhere as well. What we needed to do next was to develop a science of randomness. And we did just that. We call it probability.

Probabilistic Thinking

The fourth big idea that dawned on us around the time of Pascal and Fermat was that of thinking probabilistically. This is much more than the vague notion that some things are more likely to happen than others. Ancient writers tackled that topic with gusto, opining on the nature of coincidence, luck, unpredictability, chance, and uncertainty. Aristotle declared that "the probable is what usually happens" and riffed on that thought, contrasting it to things that are certain and things unknowable. However, this was all just philosophy, not math, and there was no attempt to turn it into a science. So we had to discover not just the idea that probabilities *could* be calculated in a systematic way, but also how to do it. The surprise we found was that probabilistic thinking can be applied to virtually every aspect of life, not just dice rolls and coin tosses.

Probability is expressed as a number between 0 and 1. By convention, we usually refer to probabilities by percentages, as in there is an 80 percent chance of rain, not a 0.8 chance. Something with a probability of 0, such as the sun rising in the west, cannot happen; conversely, something with a probability of 1 must happen. Those aren't even really probabilities. Everything between 0 and 1 is where all the action is, and that's what we began to wrap our heads around during the Renaissance. Technically, all probabilities are estimates, and they aren't something you measure, like a mile or a minute. Because they are expressed as degrees of likelihood, those who traffic in probability generally don't make predictions per se, but rather they talk about levels of confidence. I recently saw this aptly expressed in a single-panel comic where a TSA agent asked a statistician if he was carrying anything explosive in his luggage. "Probably not" was the answer and, strictly speaking, the correct one.

Nearly four hundred years in, it would be nice to say that we have the science of probability all sorted out. Not even close. In a 1950 book called *The Nature of Physical Reality*, philosopher of science Henry Margenau wrote, "General discussions of the meaning of probability by philosophers have lately shown little evidence of agreement upon any common view, and the literature is becoming progressively more confused." It hasn't gotten any better in the decades since.

What's the problem? For starters, consider the observation that there is a 0.50 chance a coin will come up heads. What does this mean, exactly? Does it mean that "as *far as I can tell*, it is equally likely to come up heads as tails"? That is, that a probability is a measure of ignorance? Or, put another way, does it mean there really is a 100 percent chance it will come up heads, but we can't measure and compute all the physical forces at play that quickly, so we just don't know? Thus there was never really any *uncertainty*, just ignorance on our part. Or does it mean that "if I flipped a coin infinity times, half the time it would come up heads"? In other words, there really is such a thing as true uncertainty.

This can all be phrased a slightly different way: Does probability exist only in our heads? Or is it out there in the universe? If the latter is the case, then what exactly is it? *Dilbert* creator Scott Adams weighs in on this in his novel *God's Debris*. A delivery guy is asked by an old man why a coin flipped a thousand times comes up heads roughly five hundred times. The old man is unsatisfied with the response, so the delivery guy asks him what the answer is. The old man tells him that there is no "why"; that all other questions have a "why," but probability uniquely does not. He elaborates:

"Probability is omnipotent and omnipresent. It influences every coin at any time in any place, instantly. It cannot be shielded or altered . . . Probability is the guiding force of everything in the universe, living or nonliving, near or far, big or small, now or anytime."

The old man is saying that probability is not in your head, not a measurement of ignorance. Rather it is the most ubiquitous force in the universe. It is in this sense that probability is the science of randomness, a science that only works because of the "miracle" we discussed earlier, that in a sufficient number of random occurrences there is an underlying order.

But there is also clearly that other sort of probability—the kind that has no independent reality and lives only in your mind. Imagine you are sitting on a jury in a murder trial. The judge has instructed you to find the defendant guilty if his guilt is proved "beyond a reasonable doubt." Interestingly, for centuries the courts have declined opportunities to put a number on what reasonable doubt means. But say in your own mind you will convict only if you are 98 percent sure he did it. You start off with no belief on the matter, then as the case unfolds, you gradually become 80 percent sure, 90 percent sure, and so on. You are certainly dealing in probabilities here, but there is no real uncertainty anywhere but in your mind. Either he did it or he didn't. There is either a 0 percent chance he did it or a 100 percent chance. You just don't know which. The probability is just your measure of your own uncertainty.

The difference between these is the difference between saying there is a one in ten chance of rain next July fourth and a one in ten chance that Bigfoot exists. At one level, they mean the same thing, and at another are completely different.

There are two types of probability. Adherents to the first, the belief that probability exists as a real thing, call themselves either objectivists, physical probabilists, or frequentists. Believers in the second type, that probability is a measure of your ignorance, refer to themselves as subjectivists, Bayesian probabilists or evidentialists. And man, they don't get along. They are like cattle farmers and sheepherders. YouTube is full of videos with each group passionately arguing their case. There's even a Bayesian/frequentist rap battle.

It is tempting to say they are both right. Maybe we are just using the same word to mean two different things and should refer to them as "risk" in the first case and "uncertainty" in the second. As Ian Hacking writes in *The Emergence of Probability:* "Students of probability, be they mathematicians or philosophers, have for centuries said that the word has two distinct meanings, and that we suffer from its ambiguity. That doesn't wash. The seemingly equivocal idea of probability seems too deeply entrenched in our ways of thinking for mere linguistic legislation to sort things out." I think he is right. Extremists on each side have a sort of religious fundamentalist view of their position as the only correct one, the only one that is scientific at all. Their cars are not likely to be sporting some sort of mathematical "coexist" bumper sticker. A more interesting debate isn't which one is "right" but which one is derived from the other.

Interestingly, both of these ways of thinking of probability came about immediately, in the mid-1600s. The distinction between them wasn't noticed, though. One can infer from the writings of Pascal, Laplace, and the rest which type they have in mind when dealing with any given problem, but they do not seem to have noticed the difference at all, easily switching back and forth between the two conceptions of probability. The subjective view, that probability is a measurement of ignorance, dominated early, as you might expect with a deterministic worldview. The chance of everything that does actually happen in the world happening was exactly 100 percent. It's just our ignorance that blurs the future. The objectivist view was of less value because there were few places to apply it outside of the uninteresting cases of dice and coins. But as more and more data became published, such as births, deaths, diseases, temperatures, and all the rest, the objectivist view became more useful.

Hacking is no doubt correct that there is a real difference here beyond semantics. Luckily for our purposes, we don't have to pick sides with either the Hatfields or the McCoys. It's philosophy they disagree on, not math, and it's the math that interests us as we explore how humans learned to see into the future.

By itself, probability is powerful. Knowing the odds of something happening is of use, but to really supercharge seeing into the future, we needed one more thing: conceiving of multiple probabilistic futures. With that, Pascal and Fermat solved the problem of points and created a new way to think about and forecast the future.

Many Possible Futures

People in the seventeenth century knew two seemingly unrelated things. First, they knew that the chance of any particular side of a die coming up in a single throw is about 0.17. That's more a static property of a die itself than a prediction. Computing the odds of rolls with additional dice and multiple tosses is more complex, but it is still the same idea. Second, people also knew that the future is full of different things that can happen. A merchant knows that his ship could sink or make it to port, and that the latter case was more likely.

There used to be a series of popular TV commercials for Reese's Peanut Butter Cups where one person is shown eating a chocolate bar and another eating from a jar of peanut butter with a spoon, and through a fortuitous accident, the chocolate lands in the peanut butter and a new taste sensation is discovered. That's what happened with our fifth and final conceptual discovery that allowed the problem of points to be solved and birthed the field of modern probability. What if you could combine those two ideas? What if various future events had probabilities that could be manipulated numerically like dice? What if you could put probabilities not just on one set of events but on each potential offshoot of those as well? And then on each potential offshoot of those, all the way to some kind of resolution?

Today we aren't wowed by this way of conceptualizing the future. It seems pretty obvious to us, but back then it sure wasn't. As Keith Devlin of Stanford University, an authority on the Pascal-Fermat correspondence, said, "The idea of predicting the future with numbers . . . means hanging numbers not on things in the world we live in but hanging numbers on tomorrow and next week . . . That was something that was just thought not to be possible."

This approach is akin to running a simulation of all possible futures, most of which will never happen. The goal is not to pick which one will come to pass, just how likely each is. Before 1654, most people had approached the problem of points by looking at the past—what had happened in the game so far. How could you do anything else? Even Aristotle didn't think you could use math to predict the future, because math is about certainty, not the uncertainty of tomorrow, when anything could happen.

On any given turn, chess players have a set of moves they can make. In their heads, they anticipate how their opponent would respond to each of them, and then how they in turn would respond to that. Even good players can see just a few moves ahead. The events of our day-to-day lives are like this, and the number of moves ahead that we can see is likewise limited. What we learned to do was to dip down into our numeric reality to see all the possible chains of events that could happen and then estimate the likelihood of each. We turned this into a science whereby we statistically model the future, not predict it.

These are the five big ideas we had to discover to solve the problem of points and invent our modern way of conceiving of the future. We discovered that our reality is numerical, and we figured out the fatal flaw in commonsense determinism: that complex systems behave chaotically. From there we discovered order in randomness, which we used to create the science of probability, and we applied that to a new understanding of possible futures.

With all of that, the problem of points all but solves itself. The reasoning might run something like this: Math would be needed to solve the problem posed by Harry and Tommy's interrupted game. The outcome could not be foretold and depended on randomness, but we could confidently assign probabilities to the random events. The number of ways the game could have played out and concluded was finite, and the probability of each outcome could be estimated. The pot should thus be split in proportion to the likelihood of each player winning the game.

Today, we still have not mastered any of those five big ideas. We have an incomplete understanding of the fundamental nature of reality, as well as the degree to which it is deterministic. The real nature of probability is still hotly contested, and we have no consensus definition of randomness or even agree whether it exists. The way we model the future is inconsistent, with competing models being championed by advocates with the passion of sports fans rooting for their team. In the 1600s, mathematicians knew much less of any of this, but they saw or intuited enough to found the science of probability, which caused the word "predict" to break off from its longtime synonyms—"foretell," "prophesize," and "divine"—to mean something quite different. What did we do with this new way to see into the future? That's our next stop.

Probability Explodes onto the World

We're about to pick our story back up where we left off in the 1650s. But before we do, just stop and imagine from that vantage point how improbable our myriad uses of probability would have seemed to them. Today, we use probability in almost every aspect of life. We don't even notice it. Whether it is explicit or not, we are running odds constantly in our head, about the likelihood of hitting traffic on a certain route home, for instance. But we also use it in our health, such as judging the efficacy of a vaccine or treatment, and in our finances, such as assessing the risk of a certain investment. At a societal level, it is used in science, transportation, government, defense, and sports. Probability powers supply chains, keeping just the right amounts of products in stock; it's in your inbox in the form of your spam filter, and all voice and image recognition programs are solely driven by probabilities. Then, underneath everything, out of sight, it performs a million other functions, such as powering the algorithms that optimize elevators in tall buildings and times on traffic lights.

But even that is just the tip of the iceberg. If you are one of the millions who are married to someone they met on a dating site, then you basically kiss a probability good night before you go to bed. A probability selects for you news items you might enjoy, matches your résumé to likely jobs, and decides whether you get a loan. Nations go to war over assessments of probabilities, such as the likelihood of belligerent states having chemical weapons or the intentions to develop nuclear ones. And it all began with the problem of points.

The details of the Pascal-Fermat correspondence became widely circulated in the close-knit world of European intellectuals, setting off tremendous interest in the new discipline of probability. Europe just happened to have a larger than normal crop of geniuses who were willing to take the probability ball and run with it. One of these was Christiaan Huygens, a Dutch mathematician, physicist, and inventor. His name no longer comes up at cocktail parties, but he is one of the greatest scientists ever. He made major discoveries in several fields, invented a compound telescope eyepiece, developed the law of centrifugal force, and discovered Saturn's moon Titan. In his spare time, he invented the pendulum clock.

Pascal and Fermat birthed probability, but Huygens raised it to adulthood. He had been interested in odds relating to gambling for a while, and after he learned of the recent Pascal-Fermat letters, he wrote a book developing the ideas in the letters into a full-fledged science. In 1656, Huygens sent the manuscript to Pascal and Fermat for thoughts and suggestions. Both men replied positively, and each included a probability problem or two to add to the set at the end of the text. The next year, the sixteen-page book, *On Reasoning in Games of Chance*, was published. The gambling angle was a great hook, and it was widely circulated.

In those pages, Huygens was quick to point out that he did not originate the ideas but had built upon the work of Pascal, Fermat, and others. This was true, but he was also being gracious. The book did develop the math tremendously. One of Huygens's innovations was the idea of expected value: a 1 in 1,000 chance of winning a thousand dollars is worth a dollar. Second nature to us, it was a useful innovation for comparing alternative courses of action. Even though he had written ostensibly a gambling book, Huygens expanded the thinking on where else probability could be applied. For the next half century, this was the go-to text for probability.

Pascal was largely done with math, opting instead for religious thought. He brought the probability mindset to that field when he created his famous rational argument for faith, Pascal's wager, which included the idea of infinity as well as decision theory—the mathematics of using risk assessments to make optimal decisions. Fermat, too, went on to make contributions in other fields, though he would occasionally return to probability. The rest of the continent went wild over probability. Not just the math itself, but all the places that it could be applied. It is a practical craft, of little use in astronomy or chemistry, though essential in matters financial, demographic, and statistical. Born of dice and cards, it had little pretense, and it was soon put to use to solve all manner of practical problems, such as the selling of annuities and the issuance of life insurance.

Even those who did not work directly on developing probability as a science were interested in how they could apply it to their field. Isaac Newton carefully annotated his copy of Huygens's book, and Gottfried Wilhelm Leibniz, a German who would create a calculus to rival Newton's, made a series of philosophical observations about the nature of probability, including thoughts on what we would later call game theory. He also figured out an innovative way to test for randomness. Say there is a piece of paper with a dozen drops of ink on it. Did they land randomly or were they placed there deliberately? He posited that the more complicated the formula you needed to produce that distribution, the more likely the dots were there randomly.

As the 1600s wound to a close, probability was to take another leap forward thanks to Swiss mathematician Jacob Bernoulli. He proved the law of large numbers. Most of us know it in its simple, intuitive form: The more times you flip the coin, the closer your observed result will be to the theoretical one. That's true, but that's not exactly what the law of large numbers says. It says that the more times you flip the coin, the higher the probability that the observed outcome will be in a specified range around the mean, and that given a desired confidence level, you can compute the number of tosses you need to achieve it.

We regularly see this today in polling. Most polls say they are accurate to within plus or minus 3 percentage points. What is usually in the fine print is that this is true only 95 percent of the time. You may have noticed that many polls are based on about a thousand people. That's because to get a 95 percent confidence that your results are within plus or minus 3 percent, you need to query just over a thousand people. Bernoulli sorted out the math on all of this. With any two of those three numbers—acceptable error, confidence, and the number of people polled—you can compute the third.

This was an invaluable tool because now we could move beyond just computing probability and add an entirely new dimension, that of sample sizes needed for levels of confidence and rates of error. The philosophical "aha moment" of all of this was the realization that while we live in a world without certainty or perfect knowledge, we can pick our acceptable level of uncertainty and error, knowing exactly how likely it is that we are wrong. That was at least something. All of this was presented in a book called *The Art of Conjecturing*. The ideas were partially disseminated by the 1690s, but Bernoulli kept fiddling with the manuscript, and it didn't come out until after his death in 1705. Three hundred years later, it is still in print.

To have been a person of science back then must have been invigorating, for everything seemed newly opened to the investigation of rational minds. You could decide you were going to be the person who figured out sound or hydraulics or the circulatory system or, well, anything, and have a reasonable chance of discovering something new. Isaac Newton stuck needles in his eye to understand color better, Santorio Santorio spent thirty years measuring everything that went into his body versus what came out in order to understand metabolism, and Jean-Antoine Nollet worked on measuring the speed of electricity by running a current through a mile of monks holding hands. Where would we apply the new science of probability? That's our next stop.

Death and Taxes

The first place we applied probability was to mortality. The data was easily available and of high financial value since it could be used to price annuities, which governments used to raise funds. In fact, the real mystery is why mortality data hadn't already been collected. It didn't require probability at all, or anything beyond simple counting. Anyone could have walked around cemeteries and recorded the ages at which people died or scoured through old government records of births and deaths. But everyone evidently had better things to do with their Sunday afternoons. Until John Graunt came along, that is.

Who is John Graunt? He wasn't a mathematician, a prodigy, or a scientist. Born in 1620, he was just a simple London haberdasher. For reasons lost to history, he began poring through the weekly bills of mortality for London that had been compiled since the late 1500s. They showed how many people died of what cause each week in London, as well as the number of births. Their primary purpose was to help the government understand the virulence and extent of the city's periodic plagues. This accounts for their sporadic publication. They are really fascinating to look through because they include roughly sixty causes of death and the number of occurrences of each, including such classics as "frightened to death" and "found dead in the street."

Evidently Graunt found them fascinating as well, for in 1662 he published an analysis of the mortality rates in London from 1604 to 1661 based on this corpus of data. And what a work it was! No one had done anything like this before. He didn't just collect data; he drew brilliant inferences from it and presented them with explanatory comments. At age forty-two, this seller of buttons and ribbons had single-handedly created the entire science of demography in one fell swoop. Just who did this guy think he was? A Swiss patent clerk?

In Graunt's time, no one knew the population of England, or even that of London. London was supposed to comprise a million people, but this was a guess and, it would turn out, a bad one, since Graunt concluded the population was well under half that. He also inferred that England and Wales must have a population of about 6.5 million, a number that we now think is a bit high but in the ballpark. Noting that there were always more deaths than births, he estimated the net number of people who moved to London each year. Clever, wasn't he? By 1669, Huygens was able to improve upon Graunt's conclusions using more advanced math, and several other European nations duplicated Graunt's work using their own local sources of data. All of this just fifteen years after the Pascal-Fermat letters.

One of Graunt's innovations was the creation of the mortality table, an estimate of the death rate at different ages. Graunt had just two pieces of data to base it on. First, he had reason to believe that mortality before age six was 36 percent. Further, he believed that only one person in a hundred would live to seventy-six or beyond. From this, he built a table that showed decade by decade how many people in a given population would die. It looks pretty compelling, and he never explained how he computed it, but if you look at it yourself for five minutes or so you can figure out exactly what he did. He just assumed a constant percentage of people alive at the beginning of the year would die regardless of their age. Put another way, the odds of a ten-year-old dying next year were the same as a thirty-year-old or a sixty-year-old. This clearly isn't how the world works. The older you get, the more likely you are to die next year. In fact, in the US, it goes up every year after age ten. By the time you are sixty-six, it crosses a threshold of 1 percent, which some see as the passage into old age. If you happen to hit one hundred, there is a 30 percent chance you will cash in your chips by your next birthday. In Graunt's time, people didn't know that and assumed that once you left infancy, your odds of dying were constant regardless of your age.

That may be hard to wrap our heads around, but people really did believe this. In our society, where premature death is rare and most people live to old age, the idea of mortality rates rising with age seems obvious. But imagine you were born in a world where few people died of old age. Someone might die at sixteen of an infection, or twenty-two in childbirth, or thirty-one from consumption (tuberculosis), or forty-seven from getting kicked in the head by a mule. Death would look quite indiscriminate in its choice of victims. If you were to ask someone whether a thirty-year-old has a greater chance of dying than a fifty-year-old, how could they know that or even have a good intuition about it? Either of them could get kicked in the head by a mule tomorrow. Or smallpox could come through and kill the young and the old without preference. So people universally thought that, aside from infants, everyone had the same chance of dying in a given year.

Well, that's what most people thought. But in Breslau, Poland, there was an ancient superstition that people aged forty-nine or sixty-three were more likely to die. A local myth-busting German pastor named Caspar Neumann decided to debunk this and collected the ages of death for his entire area using a variety of data sources as well as frequent strolls through graveyards. He showed that there was nothing special about forty-nine or sixty-three. In addition, working through Leibniz, Neumann got this data into the hands of an eager Edmond Halley, who used it to make an accurate mortality table built from real data, which he published in 1693.

The new table had an interesting implication. If life expectancy did not vary by age as was believed, then that suggests that death itself is basically random, that the Grim Reaper is a capricious fellow. But once you realize this isn't true, that there are actually discernable patterns in mortality, that changes things a bit. Death becomes a matter of probability that can be computed.

Mortality tables interested the likes of Halley not for purely scientific reasons but for pricing financial instruments such as life insurance and annuities. Annuities have been used by governments to raise funds for centuries, even as far back as ancient Rome. The deal was simple and straightforward: you pay a lump sum to the state and it pays you a fixed amount every year until your death. Knowing how to price them based on life expectancy at time of purchase was a big deal, or so one would think.

Pop quiz: Two people walk into a government office wanting to buy an annuity that will pay $1,000 a year until their death. One of the people is twenty years old, and the other is eighty. Who should be charged more for the annuity? The younger person, obviously. But before the 1650s, everyone paid the same price. Anyone could, for instance, buy that annuity for "eight years purchase," which meant that if you think of your payment to the state as a loan at the prevailing interest rate, then the state would issue to you an amount that would pay off the hypothetical loan in eight years. But since it's an annuity, the state keeps paying until you die. If you died before eight years, then the state won; it got to "borrow" your money and didn't have to fully pay you back. If you held on for more than eight years, then the state had made a bad deal. The "years purchase" multiple might vary by place or time based on the reliability of the seller and present economic conditions, but the price usually didn't vary by age. This made sense if the odds of anyone dying in a given year were the same regardless of age, which is what was widely believed but emphatically not the case. Because of this, annuities were frequently priced too low and were purchased by young, healthy people, causing real problems for governments.

How should they have been priced? A Dutchman named Johan de Witt had a key insight on this question. De Witt was an interesting character: both a gifted mathematician and a notable statesman who reached the heights of power in the Dutch Republic before he was killed and partially eaten by a pro-monarchy mob at the age of forty-six. But in 1671, just before his very, very bad day, he published a method for valuing annuities based on the age of the buyer. He combined mortality rates with Huygens's expected-value concept to figure out the likely amount that would be paid out over the life of the annuity. This was a high level of sophistication and a huge step forward, but it was not picked up by the English government, which was still happy to sell annuities for seven years purchase to all comers, young or old.

This brings us back to Halley's work from 1693. His main interest in the mortality tables wasn't just to see how likely he was to die next year. He took his much better mortality data and performed an even more sophisticated annuity pricing calculation than had de Witt, one that more accurately valued the discounted value of predicted future payments. All of this was still just forty years after Pascal and Fermat struggled with the problem of points.

The quest for ever more accurate mortality data drives the actuarial industry to this day. In part, this is because life insurance and annuities are opposite sides of the same coin. While the person who sells you life insurance wishes you a long, prosperous life full of many monthly premiums, the one who sells you your annuity is incented to suggest you take up hang gliding. If mortality tables are too pessimistic, that is, they think you are more likely to die than you are, then life insurance will be overpriced, resulting in fewer sales, while it will be a buyer's market for annuities that will bankrupt you. Obviously, if the tables are off in the other direction, the whole thing is reversed.

Laws

Tycho Brahe was a Danish astronomer in the 1500s, most famous today for the gold nose he wore, having lost his original one in a duel at college. What is often left out of that story is what the duel was over: a difference of opinion regarding a mathematical formula. That pretty much sums up the intensity of passion that was Tycho Brahe.

Perhaps he was born at the wrong time. An astronomer living before the time of telescopes had only the naked eye with which to take celestial measurements, and Brahe was a fanatic for accurate measurements. Even without magnification, he achieved a level of precision estimated to be ten times that of anyone before him. He did this by spending a lot of money inventing and building specialized equipment that was so good that there was an unfounded rumor that Johannes Kepler killed him just to get unfettered access to it. In addition to having the best equipment, he achieved such precision through sheer repetition. He would make the same reading again and again, night after night, and average them together.

The idea of averaging would seem to be pretty obvious, but it wasn't. Astronomers, including Kepler and Galileo, evidently used it, but just as a trick for reconciling different astronomical measurements, not as a general mathematical tool. Aside from astronomers, one would be hard-pressed to find anyone else averaging numbers of any kind for a century, in part because there just wasn't that much data to average. Dairy farmers weren't keeping track of average daily milk production by cow, nor were parsons taking roll in their parishes to compute the average number of the faithful who showed up each Sunday. We think like that today, but they didn't. If they had Excel, or even owned a pencil, they might have, but neither existed.

If you think about it, averages are kind of meaningless. Perhaps you have heard of the man who drowned in a river whose average depth was one foot. The idea behind an average sounds a bit perverse: "We have a hundred pieces of accurate data. Let's add it all up, divide by a hundred, write down the answer, then throw the original data away." When I tell people that Jeff Bezos and I have an average net worth of $100 billion, they seldom seem all that impressed. Statisticians get this, which is why they are more interested in how things vary from the average than the average itself. In fact, statistics can be thought of as the study of variability.

But then there's Tycho Brahe. He was able to get better data by throwing data away. Through averaging, mistakes he made one night canceled mistakes he made on other nights. A dart player may play all night without getting a bull's-eye, but the average of all of his throws probably is one.

Decades later, in 1632, Galileo published his *Dialogue Concerning the Two Chief Systems of the World—Ptolemaic and Copernican,* in which he moves beyond just taking averages. He had noticed something interesting about the errors astronomers made in their observations: small errors are more common than large ones, and the errors themselves seemed to be made equally in all directions. He suggested taking what we now call the median, in effect, lessening the impact of the outliers. This is a bigger deal than it sounds like at first, and we'll return to it shortly.

Imagine you flip a coin ten times. How many heads are there likely to be? Well, with one flip of ten, you can't really be sure, but odds are pretty good that there will be either four, five, or six. Now imagine that you repeat this ten-coin flip 1,024 times, each time writing down how many heads you get. If you were to plot those numbers on a graph, you would (almost certainly) get a familiar-looking bell-shaped curve. The peak of that curve would be at five heads, the midpoint between zero and ten. From that peak, six and four heads would put in fewer appearances than five, then seven and three even less, all the way down to zero and ten at the extreme ends.

Statistically speaking, in 1,024 flips of ten coins, you will get zero heads only once. Why? With one toss, the odds of getting tails are in 1 in 2. To get two tails in a row is 1 in 4; for three tails, 1 in 8; all the way down to ten tails—that is, zero heads—in a row, which occurs in 1 in 2^{10} times, that is, once in 1,024 times. Using this logic, how many times would you expect exactly one head? You don't actually need math to solve this problem. If there is only one way for there to be zero heads in ten tosses, there are ten ways to get one head: the first flip could be a head, followed by nine tails, or the second flip could be the only head, with the nine other tosses all tails, and so on. So how many times are you most likely to get five heads? There are 252 ways to get that. It could be HHHHHTTTTT or HHTHHT-THTT or any of the 250 other combinations that consist of five heads.

Maybe you have seen this demonstrated at a science museum with a Galton board, a flat box mounted vertically that contains a number of wooden pegs arranged like a pyramid and a hole in the top. A ping-pong ball is dropped into the hole and immediately hits the top peg. The ball can bounce to the left or the right and does so with equal preference. Having so bounced, it lands on another wooden peg, where it bounces again to the left or right, all the way down. After ten levels, the ball comes to rest at the bottom in one of eleven slots. Subsequent balls that end up at the same place will stack on top of it. If you pour 1,024 balls into the hole, they will arrange themselves on the bottom in the bell-shaped curve in what we call a normal distribution. The term bell-shaped curve is just a general term for the shape, whereas a normal distribution is the specific curve described above.

I keep on my desk a small Galton board filled with BBs. I turn it over again and again, watching the BBs dance their way to the bottom, and each time I do it, I get almost the exact same curve, the normal curve. My Galton board has 1,024 BBs, and on each flip, the number of BBs that bounce ten consecutive times to either the right or the left is quite low, usually just one, two, or three in total. There is always a big hump in the middle with hundreds of BBs. If someday I flip my Galton board over and I get a trough—high points at each end and a gulf in the middle—well, the world is about to end or something because that's just not how the universe works.

I belabor all of this to drive home a point that we will be spending a great deal of time on: The normal distribution is a product of randomness, of balls bouncing off pegs and coins flipping in the air. Every time you see it, a random process is lurking underneath it. This is strange because we are going to encounter it in places that don't seem to be driven by randomness at all.

Who was the first person to flip a bunch of coins and stumble upon the normal curve? Pride of place goes to Abraham de Moivre, a French statistician and consultant to gamblers. He was forced to live in England because of his religious beliefs, and he thereby became good friends with Newton and Halley. At the beginning of the 1700s, he noticed that binomial distributions—things like coin tosses—took on a curve shape and that the more times you flipped the coin, the smoother the curve got. He realized he could use this as a shortcut for computing the odds of getting, say, over thirty heads in fifty tosses, the sort of problem that came up in his gambling work. For small numbers of tosses, you can just count the number of possible permutations the way we did above and compute the odds from that, but with more tosses, that method becomes unworkable. So he figured out a formula that would compute the odds by determining the area under parts of the curve, a perfect application of his buddy Newton's calculus. He published his findings in 1733 in the second textbook on statistics ever written, *The Doctrine of Chances*. And there the matter rested.

You'll remember that Galileo observed that in astronomical measurements, small errors are more common than big ones, and that errors are pretty random in all directions. Over the years, other astronomers who wrestled with reconciling several slightly different readings had made the same observation, or variants thereof, but none of them ever really ran with that ball and figured out a systematic way to make accurate estimates with multiple different measurements.

Enter Pierre-Simon Laplace, arguably the most brilliant scientist France ever produced. He was born in the mid-1700s and lived for seventy-seven years, mastering a dozen sciences, producing volumes of work, advising Napoleon, and having strokes of brilliance centuries ahead of his time, such as the idea that some stars might have so much gravity that even light could not escape them, that is, black holes. He, too, was flummoxed by discrepancies in his own astronomical calculations and set about to figure out the best way to reconcile different measurements. He solved this problem in 1778 by creating what we call the law of errors, that is, that if you graph all your different measurements of the same phenomenon, you get a curve just like de Moivre's, a bell curve with a normal distribution.

At first, this doesn't seem to make any sense. De Moivre got his curve by flipping coins; his graph is a graph of randomness. Laplace's graph was one of errors, a very different thing. But upon reflection, the common thread emerges. Each stellar observation is an independent event, unaffected by the other ones. And each one is subject to a range of potential problems such as distortions in lenses, differences in timekeeping accuracy, variations in atmospheric conditions, and so forth. Think of each of these factors as a peg in a Galton board. Sometimes, often in fact, the inaccuracies offset each other and the observation turns out to be pretty accurate, the way the ping-pong ball bounces to the left and right a few times each and ends up right in the middle. Other times, much less often, all the errors are in one direction, and you get a measurement that is way off. In this way, Galileo's two observations—that small errors are more common than large ones and that errors are equally likely in either direction—give us our symmetrical normal curve in which the hump in the middle is all the measurements that are just a little off. The very top of the curve, which is the mean, the median, and the mode, is the best estimate of the true measurement.

Fast-forward to 1809, and we find Laplace publishing a proof of what we call the central limit theorem. It's an important idea that merits exploration:

Galton boxes give you a normal distribution because the balls randomly bounce to one side or the other as they go down. Normal distributions are handy things because if a set of data is distributed that way, then all kinds of interesting calculations can easily be done on it relating to standard deviation, variance, and so forth. Put simply, statisticians love normal curves because we know a lot about how they work. But what of things that don't fit on a normal curve? What about data that is distributed in other ways? Are we precluded from understanding much about that? Not with the central limit theorem.

Say you work at a candy manufacturing company. The chocolates are supposed to be exactly the same weight, but they aren't, for a number of unrelated reasons having to do with the machinery and the ingredients. The machines make a few that are perfect, then a few that are way off in one direction, then a couple that are too light. Then the machines are recalibrated, and for a while the chocolates are pretty consistent, then too heavy. Then just a little too heavy, and so forth. All over the map. We don't have our nice, clean metal balls bouncing to the left and the right with equal affinity. We have this hodgepodge of candies whose weights are not a normal distribution. Perhaps a graph of their weights is wavy and goes up and down. Or it has two humps in the middle, or maybe it is a trough. What do we do then to understand the distributions of our candy's weight?

Let's say you work on the assembly line at the factory and are responsible for packing chocolates in boxes of thirty. The central limit theorem says that the weight of the filled *boxes* of chocolates will follow a normal distribution, even if the weight of the individual chocolates doesn't. Crazy, isn't it? The weight of the chocolates isn't normally distributed, but the weight of the boxes of thirty is.* This is important because it means data that itself isn't normally distributed can be made to behave that way by, in effect, putting several randomly selected pieces in a box of chocolates.

This was all on Laplace's mind in 1810 when he stumbled upon a new book called *Theoria Motus* by German mathematician Carl Friedrich Gauss. Gauss had performed a minor miracle of the day. For a month, Italian astronomer Giuseppe Piazzi had been tracking a dwarf planet he had discovered called Ceres. It then disappeared behind the sun. Piazzi waited until it should have emerged but couldn't locate it. Various mathematicians and

---

* Thirty is the kind of magic number where the central limit theorem really kicks in. Boxes of three, for instance, aren't much closer to the normal distribution than boxes of one.

astronomers competed to use Piazzi's data to help him find his lost planet. With precious little data to work with, Gauss spent three months calculating where it was and predicted a location pretty far from where everyone was looking. He turned out to have been right, and his coordinates were accurate to half a degree. In *Theoria Motus* he explained how he did it, using a method of least squares, or what we today call regression.

Laplace saw that this was another manifestation of the normal distribution that had considerable overlap with his theory of errors and central limit theorem, and he used it to advance his own work. Gauss, too, continued to develop principles around the normal distribution, which today is often referred to as a Gaussian distribution. That isn't a very good name for it because while Gauss had been a child prodigy, having corrected a math error of his father's at age three, he was only two when Laplace published his law of errors, so it is doubtful that they had collaborated. Gauss and Laplace referred to the normal curve by a variety of names, including *law of deviation* and *law of frequency of errors*. It was Francis Galton who much later popularized the phrase Gaussian distribution.

All of this would largely be mathematical arcana if not for the fact that so many things in biology, sociology, and psychology were found to be distributed along normal curves. This is a bit troubling because the normal curve is the product of randomness, suggesting that we are as well.

Data

In 1729 the government of France held a lottery to raise funds. Voltaire read the fine print of the convoluted rules and, with the help of young mathematician Charles-Marie de la Condamine, determined that the lottery could be legally gamed in such a way that with a small amount of money he could guarantee himself a fortune. To put it in the language of our time, say the government wants to encourage people to buy US savings bonds. So it says, "Buy a bond for any amount and we will throw in a lottery ticket. If you win the lottery, we double the value of your bond." This would work fine. People who bought a $1,000 bond might win $1,000; people who bought a one-dollar bond might win one dollar. However, what the Paris government did was in effect say, "If you win, not only will we double the value of your bond, but we will give you a million dollars, too." Well, that changes things, doesn't it? Voltaire realized all he had to do was buy huge numbers of ridiculously small bonds so that he was all but guaranteed to win the value of his bond (one dollar) plus the million dollars. So that's what he did. He later modestly wrote that he "got lucky," but the truth was that he "got mathy." The government cried foul when it figured out what he had done, but since Voltaire had broken no rules, it fired the minister of finance who had dreamed up the whole thing.

Yeah, I know. Today, everyone would figure out how to game that. That's child's play. I tell this story to show the degree to which sophistication on matters of math and probability was still lacking in governments at that time. One can argue it is lacking today, but that's a topic for another book. Governments were often doing things like this with perverse incentives, inadvertently creating Ponzi schemes, disastrous debt instruments, and manias of all sorts.

It wasn't just governments that lacked sophistication. They just happened to operate at a scale that amplified the effects of their poor computational skills. But as the century progressed, so did our proficiency with probability and data. In 1761, Reverend Thomas Bayes proved Bayes's theorem, an embodiment of the subjectivist view of probability whereby your calculations of the probability of an event change as you acquire more information. This is the approach that one might informally use while on the jury in our aforementioned murder trial. It allows you to constantly update your own belief in the guilt or innocence of the accused as new evidence is presented. Bayes formalized exactly how to do that mathematically, and today his name is spoken with a kind of hushed reverence by subjectivists.

In addition to advancements in the math behind probability, the eighteenth century saw great strides forward in how data is presented, making it more accessible to a broader group of the educated public. In 1735, mathematician Leonhard Euler was curious how someone could hypothetically cross each of the seven bridges into the town of Königsberg just once and determined that it was impossible, creating what we today call graph theory. In 1752 the first contour map was drawn, and notation for 3D space, with x, y, and z coordinates, was invented. A year later we began visualizing data with timelines, followed shortly by the color coding of data. A few years after that, we invented the line graph; *Playfair's Commercial and Political Atlas* gave the world the bar graph in 1786; and eight years later the first commercial graph paper came out. The development of the pie chart closed out the century.

With greater sophistication in both our mathematical techniques and the presentation of their results, the only thing lacking was some data we could apply all of this to. But that was coming as well. By the late 1700s, a kind of data collection known as phenology was all the rage among the intelligentsia. Phenology is the study of periodic events in biology, such as the appearance of the first butterflies or certain fruits. It is not a new thing, as the Japanese have been keeping track of the first appearance of the cherry blossom for over a millennium, but it became much more widespread in the late eighteenth century. Thomas Jefferson was a fanatic about it, compiling volume upon volume of original data over the course of decades. For instance, you can see among his papers a vegetable market report for 1801–1808 that he compiled *while he was president of the United States* that shows the dates that roughly forty fruits and vegetables were for sale in the markets of Washington. It is even rendered in a timeline of the sort that had just been invented. All over the world there were naturalists like Jefferson keeping track of everything, including daily temperatures, the first snowfall, the migration of birds, the laying habits of chickens, and much more.

At the same time, governments were getting into the data game in a big way. Data goes hand-in-hand with bureaucracy, central government, and industrialization. Agrarian economies without standing armies had little interest in demographics, but the modern state is driven by data. In fact, our word "statistics" originated at this time from the German *statistik*, meaning the information needed to run a government.

Government data was systematically collected and processed at great scale, and much of it was about people and society. Governments counted births, deaths, illnesses, murders, suicides, hospitalizations, bankruptcies, lawsuits, and much more, frequently in multiple dimensions, such as illnesses by occupation and neighborhood. They would record every major crime in great detail, such as the weapons used in each murder. Suicides were of particular interest, and the data of each one was carefully recorded along with the time of day, method, gender, age, marital status, and religious affiliation.

By 1830, enough data had been collected that people noticed something truly shocking: these numbers that they were compiling had amazing regularities that seemed almost unbelievable. From year to year, or even month to month, there was a basic stability to many of them. This had been noticed long before in the ratio of boys to girls born each year, but who would have guessed there would be similar stable ratios for murders, right down to how often each weapon is used? How could that be? The Almighty may hold gender ratios in check for our well-being, but why hold constant the weapon choice in murder as well? Even today, it is hard to wrap our minds around these sorts of statistical consistencies. The number of people who died in the US in car crashes in 2018 was 36,500, while in 2019 it was 36,100. Likewise, in the same two years, the numbers of workplace deaths by electricity were 160 and 166, respectively.

How can this be? It's the Galton board at work. Not all normal curves have to be binomial, that is, two equally likely options, heads and tails. Roll a twenty-sided die a hundred times and count how many times you roll a thirteen. On average, in one hundred rolls, you will throw a thirteen five times. But sometimes it will be more than five times or less than five times. Repeat the one hundred rolls enough times, graphing the thirteens, and you will get a normal curve.

Think of each time someone drives a mile as a roll of four one-hundred-sided dice. It is reasonable to assume that the total number of rolls (miles driven) each year is about the same; we have the same population, the same number of workdays, and the same number of Friday nights. Fatal car accidents are the results of multiple unlikely adverse events all happening. Someone has to run a red light, then someone else has to not see them, then they have to swerve into oncoming traffic, there has to be a car right there, and so forth. If each of our hundred-sided dice has a black X on only one face to signify an unlikely bad event, and it takes four of those at once to result in a fatality, then in every hundred million rolls of four dice there would be one time they came up with four black Xs. And that is in fact the fatality rate in the US: one death for every hundred million miles driven. To get to the 36,100 fatalities of 2019 requires 3.61 trillion rolls of four dice. That is the total number of miles driven in the US that year.

Viewed this way, one can account for stable statistical averages for automobile deaths. They can shift over time, of course, as cars get better and safer, but from year to year they are stable. How far can you push this Galton-board view of life? Does it account for differences in human behavior? Do we say that the decision to commit a murder is likewise the roll of a handful of many-faced dice that will compel you to murder someone if they all come up a certain way? Lambert Adolphe Quetelet sure thought so.

Is That Normal?

Quetelet (pronounced kate-lay) was a Belgian mathematician born in 1796. He started out as an astronomer, and his hero was Isaac Newton. He was amazed at how Newton had found simple laws that explained so many phenomena. Wanting to follow in his footsteps, Quetelet persuaded the government to build its first observatory and put him in charge of it. While he was out of the country touring other observatories, a revolution in Belgium effectively ended his dreams. Unable to return, he stewed and cursed his bad luck. Why were there revolutions? Why is society so unstable? If only someone could come up with Newton-like statistical laws that explained all that. "Wait a minute," he probably thought, "I'm someone."

Thus, he set about to create "social physics," applying statistics to humans and society. Since he was an astronomer, the tools he was most versed in were astronomical. "How different," he probably thought, "could people and planets *really* be?" So he went looking for useful data about people, such as height, weight, age, number of births—anything. He even invented our modern measurement of BMI (body mass index) along the way. The more data he looked at, the more he saw normal distributions in it all.

Consider the chests of Scottish soldiers. Quetelet discovered a medical journal that contained the chest measurements of nearly six thousand Scottish soldiers. The first calculation he performed was to determine the average, which turned out to be 39.75 inches. This simple action was revolutionary. People had averaged positions of stars before, which makes a kind of sense since we are talking about multiple measurements of the same thing. But what does an average chest size really mean? Next, he graphed all the measurements and found that they did in fact conform quite well to a normal distribution with the top point of that, the apex, being 39.75 inches. To Quetelet, this meant something. Just as the average of the astronomical observations is the true one, so was 39.75 inches the measurement of the true chest, the Platonic form, the one from which all the rest of us are errors. In fact, he likens it to errors, saying that it was as if a sloppy tailor took six thousand measurements of the same soldier. Thus 39.75 inches is the chest measurement of what he termed "*l'homme moyen*," the average man.

Many societies have the notion of an average Joe. Throughout Latin America, he is Juan Pérez; in Italy he is Mario Rossi; and in Denmark he is Morten Menigmand (Morton Everyman). Those are largely caricatures. Quetelet is not talking about this, a Jacques Six-Pack; rather he is talking about something more mathematical, like the suburban family with 2.2 kids. To understand this, he invites us to imagine a gladiator and a thousand sculptors, each commissioned to carve the gladiator in stone. No statue will be a perfect copy, but the averages of all the statues will be a perfect copy of the gladiator. Some of the "errors" in the statues might have actually made for a better gladiator if he was, for instance, given bigger muscles. But the original gladiator is still literally the "average man."

In 1835 Quetelet presented this view in *A Treatise on Man and the Development of His Faculties*. In his world, average isn't mediocre; it is ideal. It is a view influenced not only by astronomy but also by the Industrial Revolution, which also sought perfect copies of an original model, in which all variance was error.

It is unclear how far he took this. On the one hand, he remarks that "everything differing from the average man's proportions and condition would constitute deformity and disease." He says that in any epoch, if there was ever a perfect copy of the average man of this age, he would be the most beautiful person alive. In 1942, sculptor Abram Belskie put this to a test and crafted *Norma* and *Normman*, two statues of human forms made from the average measurements of over ten thousand young adults, making, in essence, *l'homme moyen* et *la femme moyenne*. A subsequent search couldn't find a human who exactly matched either of them.

On the other hand, Quetelet believed that positive characteristics such as moral goodness are also distributed on a normal curve, and it is hard to see how a person with average morals is superior to one with above-average morals.

The word "normal" in the normal distribution is unfortunate. De Moivre meant nothing pejorative; he was just flipping coins and talking about normal outcomes. But when you apply this to people, if there is a normal, there must be an abnormal. "Deviation" is a perfectly benign math term, but its human form, "deviant," isn't found on many résumés. The idea that deviation from normal is bad is still with us in innumerable ways. When you get blood work done, your scores are shown next to a reference range. The reference range is not a medical opinion but a mathematical calculation of the averages of the population. Your body may have more X or less Y than average and be just fine, but you still get the little asterisk next to your number because it isn't "normal." Referring to anything as "abnormal" is almost always negative. We never refer to a beautiful model as having an abnormal appearance.

Regardless of what Quetelet intended, fifty years after *A Treatise on Man,* his work would be cited by Francis Galton, creator of our Galton board, when he coined the term "eugenics" and propounded its tenets. We will return to that topic.

But what of the underlying idea, that the range of human traits and behaviors is normally distributed? Quetelet thought he saw normal curves everywhere. Was he right? Granted, of all the distributions in the universe, the normal one is the most common. And, using sampling, averaging, and the central limit theorem, you can legitimately make things that aren't themselves normally distributed conform to that curve. But attempts to apply it in all of the places that Quetelet did required a great deal of imagination and a bit of squinting. During his lifetime, there was even a term coined for this view of everything as normally distributed: Quetelismus.

Quetelet would periodically announce to the world that he had found yet another trait so distributed. One of these was the suicide rate, which he claimed was largely constant year by year, as were the methods employed. This is a shocking finding, because according to Quetelet's worldview, we are all—each of us—just one unlucky roll of dice away from ending our lives.

After Quetelet published *A Treatise on Man*, people started seeing normal curves all around them for attributes such as artistic ability and intelligence. Some saw it in the size of loaves of bread, the quality of wine, and the length of women's hair. Others observed it in the behavior of plants and animals, allegedly strengthening the case that it also applied to human society. Those who credited Quetelet as inspiration for their own work include Karl Marx, Florence Nightingale, physicist James Maxwell, and physician John Snow, who applied his techniques to disease and helped end a cholera epidemic. As Todd Rose wrote for the *Atlantic*, "Wilhelm Wundt, the father of exper-imental psychology, read Quetelet and proclaimed, 'It can be stated without exaggeration that more psychology can be learned from statistical averages than from all philosophers, except Aristotle.'" Of course, except Aristotle.

Quetelet saw normal curves everywhere because that was the only curve in town. No one had developed mathematical tests for normal-ness yet, so it was hard to rebuke his assertions that normal curves were everywhere. If he is right, the implication is that individual behavior and society at large are the way they are because of countless coin tosses, each completely random. It is a belief that every one of us is the composite of a million flips of the Galton board. The first flip gave you your genes, the second flip your fam-ily, the third, your circumstances. Then every day, the board keeps flipping again and again.

I personally think this is a pretty bleak view of life. It is the old deter-minism returning in new cloth, this time a kind of demographic determin-ism. Individuals have no free will, and society has a specific destiny in the shape of a bell curve. Quetelet saw all of this, writing, "The moral order falls in the domain of statistics," then adding that this is "a discouraging fact for those who believe in the perfectibility of human nature. It seems as if free will exists only in theory."

Bleak view or not, it has been neither proven nor disproven, and so remains an open question. We can discuss the implications if it is true: The

Eugenics

L et's turn our gaze now to Francis Galton, a British polymath who was born in 1822 and lived to nearly ninety. He was the grandson of Erasmus Darwin, one of the most famous men of science of his day, as well as the cousin, contemporary, and friend of Charles Darwin.

Galton loved to count things and advised others to do so as well. He counted everything, from mustaches to dog breeds, and all things in between. If he was at a lecture, he counted coughs; if at a funeral, sneezes; and if he was in a sauna, he counted . . . well, you get the idea.

He was the furthest thing from a nerd, though. He was fearless, once walking into a lion's cage without even a pause. He explored great parts of Africa where he was the first Caucasian any of the locals had ever seen. It helps to picture him as he really was, the quintessential caricature of a wealthy, gentrified British man at a time when the Empire was perched atop the world and yet still managed to have an exaggerated sense of its own importance. My favorite Galton story is when he caused offense in an African village by refusing to participate in a ritual where his host was supposed to gargle some liquid and spit it into his face. I can't help but picture him, indignant and shocked, saying, "The hell I will," when the procedure was explained to him.

His contribution to the main arc of our story is that he invented the idea of regression to the mean. It is an important concept, yet so many people are unaware of it. What it says is that if you are super smart or really tall, your children aren't likely going be as smart as you or as tall as you. Why is that?

If you are extremely intelligent (as is suggested by your excellent choice of reading material), it's because your ping-pong ball bounced to the "smart side" ten times out of ten as it went through the Galton board. You got really lucky. Now, we take your ping-pong ball from the box, pretend it is your child, and drop it back in. It is unlikely to bounce ten times to the "smart side." It might bounce once on that side for "smart genes" and once more for "got read to a lot as a child," but for the eight other bounces, well, it's on its own, and odds are it won't luck out and get a perfect score like you did. As Peter Bernstein writes, regression to the mean "explains why pride goeth before a fall and why clouds tend to have silver linings. Whenever we make any decision based on the expectation that matters will return to 'normal,' we are employing the notion of regression to the mean."

An illustration of why it matters is as follows: A teacher administers a test. Half the students do well and half do poorly. The teacher believes that if the kids who did poorly just had more homework, they would do better, so they are given more homework and tested again. This time they do better! Thus, the teacher incorrectly reasons, more homework must be good.

The logical flaw here is easy to see: Some of those "bad" students were ill the day of the first test, or didn't understand it, or just performed worse than they normally do. When they are retested, it's like pouring the ping-pong balls back in at the top of the Galton board. Some bad students will seem to do better, and some good ones worse, but they are all just regressing to the mean.

Galton also invented the mathematical notion of correlation and a way to compute it. His young protégé and enthusiastic hagiographer, Karl Pearson, fully developed the notion in creating the correlation coefficient. Pearson also formalized standard deviation—a useful concept for many reasons, not the least of which is that it gives us the empirical rule specifying that in a normal distribution, 68 percent of data is within one standard deviation, 95 percent is within two, and 99.7 percent is within three.

With that, we come to the end of the 1800s, and we have in place the major statistical tools that we use today. As the 1900s dawned, one might reasonably have supposed that the new century would be one of scientific enlightenment. It was, sort of. Science continued to make great strides that uncovered mysteries of the universe and bettered people's day-to-day lives. But the century also began with a puzzling rise in spiritualism. The Ouija board came along in 1901, and it purportedly named itself, choosing the words for "yes" in French and German. According to the box of the current Hasbro edition, it is for ages eight and up, meaning that while you need to be twenty-one to order a beer, you only need to be eight to summon the Devil. Sir Arthur Conan Doyle should have been the very epitome of the same rationalism he imbued Sherlock Holmes with, but he wasn't. In 1917 he was convinced that a photo of fairies taken by two young girls was real. He would later have a falling out with his friend Harry Houdini because the latter kept debunking mediums whom Doyle believed in. And strangely enough, the "science" of phrenology was still widely believed. It would die out early in the century but remains in our vocabulary today when we describe art or humor as highbrow and lowbrow.

While Ouija boards, cranial bumps, and mysticism are largely benign, there was a new practice that was developed in the name of humanistic enlightenment and science. It was called eugenics, from the Greek words for good birth, and Francis Galton was its champion.

It started out that he was just really interested in geniuses and wondered if the trait ran in families. He determined that it did and concluded that biology was the chief reason, stating that he objected to "pretensions of natural equality." He also noticed something that should have given him pause—that genius in families petered out quite quickly. Only about a third of the children of geniuses were in that category, then only a third of their children. It is almost as if they were, you know, regressing to the mean.

Galton didn't see it that way. Rather he saw it more as their tendencies to marry heiresses. What's the connection? Because, according to Bernstein,

Galton argued that "heiresses must come from infertile families . . . [I]f they had had a large number of siblings with whom to share the family wealth, they would not have inherited enough to be classified as heiresses." In other words, families of geniuses would die out because they married infertile women. It's pretty poor logic, but it evidently satisfied Galton.

Galton proposed that the best people breed with the other best people, while discouraging everyone else from reproducing at all. In such a manner, the species could be improved. Again, regression to the mean—*an idea he invented*—says it can't be done. But intellectual blinders being what they are, he sallied forth.

As Adam Cohen, author of *Imbeciles*, told NPR, eugenics "really derived a lot from Darwinian ideas. The eugenicists looked at evolution and survival of the fittest as Darwin was describing it. And they believed, 'we can help nature along if we just plan who reproduces and who doesn't reproduce.'"

Even if one were so inclined, there is no way to sugarcoat Galton's views—they were awful. In 1904 he wrote, "Eugenics is the science which deals with all influences that improve and develop the inborn qualities of a race. But what is meant by improvement? We must leave morals as far as possible out of the discussion on account of the almost hopeless difficulties they raise as to whether a character as a whole is good or bad."

So far so bad. It is never a good sign when someone starts out with "We must leave morals out of the discussion." But that idea is at the core of being a statistocrat.

He went on to say that three things needed to be done. First, people needed to understand the "science" of eugenics and accept it as fact. Second, a method for its "practical development" had to be figured out. Cue chilling music. And finally, "it must be introduced into the national conscience, like a new religion. It has, indeed, strong claims to become an orthodox religious tenet of the future, for eugenics cooperates with the workings of nature by ensuring that humanity shall be represented by the fittest races. What nature does blindly, slowly, and ruthlessly, man may do providently, quickly, and kindly."

As you know, it didn't work out so kindly. To be clear, Galton was no Nazi, and that's the tragedy: He really meant well. Statistocrats usually do.

Three years after he wrote that, the first forced sterilization laws were passed in the US. Eventually thirty states would do likewise. The question of their constitutionality made its way to the US Supreme Court, which upheld Virginia's forced sterilization law in 1927. It was no 5–4 squeaker; the justices ruled 8–1, the sole dissenter being the only Catholic on the court, Pierce Butler, who wrote no dissenting opinion. History would do that for him.

We do have the majority opinion, penned by none other than Oliver Wendell Holmes, who lived such a long, influential life that he shook hands with both John Quincy Adams and John F. Kennedy. His intelligence was so admired at the time that Arthur Conan Doyle named his great detective after him.

This book is about seeing the future, and that's what Holmes references in his opinion. When justifying forced sterilization, he wrote, "It is better for all the world, if, instead of waiting to execute degenerate offspring for crime or to let them starve for their imbecility, society can prevent those who are manifestly unfit from continuing their kind. The principle that sustains compulsory vaccination is broad enough to cover cutting the Fallopian tubes. Three generations of imbeciles are enough."

Holmes thought he could see the future of the "degenerate offspring" so clearly that he justified not allowing them to live. By this reasoning, forced sterilization became the law of the land, and over sixty thousand procedures were performed, the last one in 1981.

Starting in the 1950s, a similar drama would play out with lobotomies (Worst. Nobel. Ever.) being performed on a wide range of unwilling patients, a practice also justified via data and concern for the greater good. The recency of this example suggests it is unlikely that the technocratic urge to perfect society through the culling of "undesirables" has left us, or perhaps even much abated.

The Next Big Thing

The first transistor computer was called TRADIC, the Transistor Digital Computer, a name only an engineer would come up with. It was built for the US Air Force by Bell Labs in 1954, the watershed year that will be our launching point for the next section. During the three hundred years from the Pascal-Fermat letters in 1654 to that computer in 1954, we steadily built out the science of probability using just our intellect, that is, without the aid of machines. We began to live in a kind of perpetual future, our gaze always on tomorrow.

It is no coincidence, I think, that the genre of futurist fiction was born during the time of Pascal and Fermat. In his book *Origins of Futuristic Fiction*, Paul K. Alkon writes that "until the eighteenth century . . . only two earlier works of this kind are known: Francis Cheynell's six-page pamphlet . . . published in 1644, *Aulicus his dream, of the Kings sudden comming to London*; and Jacques Guttin's incomplete romance of 1659, *Épigone, histoire de siècle futur*." Before this time, he argued, the impossibility of writing about the future was universally assumed.

Three hundred years later, the world was obsessed with the future. In 1954 everyone was imagining the car of tomorrow and the house of tomorrow. Futuristic fiction was everywhere, from high literature such as *1984* and *Fahrenheit 451* all the way down to pulp paperbacks with lurid covers and tantalizing titles. There were futuristic radio shows, comic books, movies, and TV shows. The word "futurist" in its modern sense was born then.

Google maintains a tool called the Books Ngram Viewer. You can enter a word or phrase, and it will show you how often it was used in published books, year by year. I set the range to 1654 to 1954 and did six queries using the words "predict," "future," "forecast," "estimate," "speculate," and "projection." In each case, the graph started at nearly zero at the beginning of the period and steadily rose for three centuries.

Nevertheless, humans have five basic limitations that continue to impair our ability to look into the future using the science of probability, and by 1954, we were feeling them acutely. Let's look at each in detail.

First, most humans aren't very good at math, especially what we called story problems back in grade school. This can be illustrated with any number of straightforward problems. Consider this: There is a disease that one person in a thousand has. A test for the disease has a 5 percent false positive rate. What is the probability that a person who tested positive for the disease in fact has it? The answer is about 2 percent. If a thousand people take the test, one will have the disease and test positive while fifty others will get false positives. Only about a quarter of people can answer questions like this correctly. Many doctors miss it as well.

Or this one. Pluto travels about four billion miles each time it goes around the sun. You get a rope that long and place it out into space along Pluto's orbit. So you have this giant, giant circle of rope, and you decide you want to expand it by a foot in all directions. How much rope do you need to add? A million miles? A thousand miles? No, just six feet. You are only increasing the radius of the circle by a foot, and the formula for circumference is $2\pi r$, so you only need two pi feet of rope.

The low rates of success with these questions are generally used by those who pose them to collectively shame us into thinking we are a bit dense. This is unfair. We may inhabit a numerical reality, but our perception of the world is not quantitative, it is qualitative. Numbers are our second language, and many aren't terribly fluent in it.

Our second limitation, even with accurate data, is that our reasoning ability is inherently flawed. For some reason, our brains have hundreds of cognitive biases, places where our reasoning isn't accurate. Why we developed these is a matter of speculation. Are they bugs in our genetic code? Or do they convey some non-obvious survival advantage? Is it because our brains are wired for stories, not logic? Whatever the case, reading a list of them is pretty embarrassing. Some examples:

- Default effect: When presented with multiple choices, we favor the default one.
- Framing effect: We come to different conclusions depending on how the same set of information is presented to us.
- Wishful thinking bias: We overestimate the chances of positive outcomes.
- Confirmation bias: We seek out data that confirms what we already believe.
- Continued influence effect: Information we no longer believe to be true still influences our decision-making.
- And, my personal favorite, the rhyme-as-reason effect: If something rhymes, it is more likely true.

Third, we are highly limited in our ability to collect and store data. To begin with, data collection is often labor intensive, tedious, and error prone. Our efforts to store and retrieve it tend to scale quite poorly, becoming exponentially more complex as they grow in size linearly. Our minds aren't built to keep clear counts of all the various events that befall us, our actions, and the outcomes; so we throw out most data and replace it with an axiom. I have undoubtedly hit my thumb with a hammer more than once, but I don't recall any specific instance, rather just that it should be avoided. This is cognitively efficient, but as data storage devices, we are woefully lacking. Additionally, all the things people learn in their lives are lost upon their death, except for the bit they pass along. Thus our collective memory as a species advances much more slowly than our collective experiences could, in theory, enable it to.

Fourth, even when we have data, our brains are limited in using it. We can only hold a tiny bit "in memory" at any one time, yet our incredibly robust sensors flood our brains with vastly more data than we can process. We are thus constantly basing decisions on partial information. We have compensated for this shortcoming by developing heuristics—rules of thumb—to deal with our limited information. They aren't cognitive biases per se because they are sort of rational. Whereas there is no reason to believe that something that rhymes is true, you can see how our tendency to form opinions with little information makes at least some sense, lest we be mired in indecision when a bear is running at us. Further, our tendency to hold on to beliefs even in the face of evidence to the contrary prevents us from constantly flipping back and forth between two courses of action as more data is revealed, such as changing our mind on whether to run away or climb a tree. The heuristic is not the best strategy, but it's probably better than no strategy at all.

We have a hard time shaking heuristics even when they are incongruous to our modern ethics. The tendency to trust people who look like you over those who don't must have made sense in a prehistoric world where kinship ties were the strongest forms of loyalty, but today it's just racism. Not all heuristics are innate, of course. We develop new ones all the time based on experience. The quality of the heuristic is usually determined by the immediacy of the feedback loop upon which it is based. Cooks learn good heuristics, while high school guidance counselors do not.

Our fifth limitation is that we are not mentally equipped to handle the complexity of how events unfold. If history has one overriding lesson, it is that everything has unintended consequences: Columbus ended the Renaissance by reorienting the world away from the Mediterranean toward the Atlantic, Gutenberg set off a series of events that led to the Protestant Reformation, and the Versailles Treaty led to World War II. Our inventions change us in ways no one ever guessed: air-conditioning undermined communities since people stopped sitting on their porches visiting with their neighbors, while the light bulb created the night shift, a way to double factory output. The internet and smartphones have had more unintended consequences than can be counted.

It turns out that the synchronists were right, at least to some degree. There are inscrutable lines of cause and effect that connect things that look unconnected. The future unfolds in a manner beyond our ability to predict except in relatively general, straightforward ways. If next month's weather is governed by chaos, how much more so is next year's election?

These are the problems innate in humans that limit our ability to see the future clearly. Can we fix them? Debug our faulty intellects? Not a chance. There are reasons we are the way we are; we are optimized for other purposes, not the least of which is thinking in stories, not logic. So we did something else instead: we taught rocks how to think.

Act III

Rocks That Think

Progress

Your body runs on about a hundred watts of power. Roughly a quarter of that is needed to power your brain, which is about a millionth of what a supercomputer requires to do far less. Your mother was right: you really are amazing.

Until ten thousand years ago, the dawn of the agricultural age, we couldn't increase our energy use at all. But that changed when we began farming, because we started using animals to pull plows, pump water, and run mills.* We only had animal power from then to the 1800s, when we got steam power by burning wood, then coal. Coal was a miracle, for it was made out of the sunlight that had fallen on the earth a hundred million years ago. Electric power came soon after, generated from what seemed like limitless reserves of a range of energy-dense fossil fuels. Today, the average Westerner uses a constant ten thousand watts of power, a hundredfold increase over our biological limit.

That's what technology does. It amplifies what people can do. All prosperity flows from it. Without technology, we would scratch out a precarious existence in which survival would be a full-time job, and life, in the words

---

* Interestingly, treadmills were invented and first used in British prisons to punish people, and their treading on them would run a literal mill.

of Thomas Hobbes, would be "solitary, poor, nasty, brutish, and short." But with technology we have abundance and ease. I do not work harder than my great-grandparents did, but I live a much more lavish life than they ever did because I use technology to multiply what I am able to do. They hauled water from a well. I just turn on the faucet.

But it isn't technology itself that entirely supports our standard of living. Consider animals. They have technology, albeit extremely rudimentary technology. Their tools are at the level of using a shoot of grass to get at some termites. Our tools are, broadly speaking, billions of times more advanced. And yet, we aren't billions of times smarter than they are. We are, to make up an arbitrary number here, a thousand times smarter than the smartest animal. There is a disconnect between our great cities and all that is within them and the lives of, say, dolphins. I'm not expecting them to have invented the internet, or even the telegraph, but you would think that after a million years they would have something to show for their supposed intellect, yet they don't. Today's dolphin lives the same exact life that the last thousand generations lived.

What's the difference? Let's say you were dropped onto a desert island with just the clothes on your back. You would immediately feel the limit of having just a hundred watts to work with, but you would likely start channeling your inner Robinson Crusoe and set to building a shelter, finding food and water, making fire, and all the rest. You would likely pat yourself on the back for all those *MythBusters* episodes you watched, and even some of the *Gilligan's Island*s. You would be able to accomplish all of this not because you figured it all out yourself, but because within you is accumulated knowledge that spans back in an unbroken lineage to the cave painters of Chauvet and Lubang Jeriji Saléh.

A real Tarzan, a baby raised by apes with no knowledge of human culture, wouldn't be able to do anything like that. Sure, he would have a mental language, but he would have to start from square one technologically speaking. Our world is what it is because of our accumulation of knowledge, not our individual intellects. A smartphone probably has an aggregate of a hundred billion years' worth of progress wrapped up in it. That number is a

guess, of course, but I doubt it is an exaggeration. We learned how to mine
the nickel for it three thousand years ago, and no one has had to discover
that again. Over those millennia, we learned how to better refine and assay
that and each of the other metals well. To accomplish this, we had to master forges and build a range of specialized equipment. We had to develop a
supply chain and the means of transportation within it, and the fuel to run
it, along with the monetary systems, the accounting systems, the laws and
legal codes around it, and, well, everything. To do that, we had to invent the
educational system, train teachers, build libraries. But to do that, we had
to invent the alphabet, master printing, invent architecture, and construct
civilization. To keep everyone working on all this in good health, we had to
invent medicine and the tools needed to practice it. To do that, we needed
farmers to grow food to feed everyone. The farmers, in turn, needed clothing
and shelter. All of that effort, the aggregate of all those person-years of work,
was needed to make a smartphone. So is the idea that a smartphone is the
product of a hundred billion years of accumulated progress all that crazy?

That's the difference between us and the dolphins. They don't have
progress. Progress requires the ability to recall the past and imagine the
future, which animals don't have or do in any meaningful way. We envision
the future and understand our agency to make it different from the present.
We developed a way for our knowledge to accumulate over the eons, with
each generation taking what we knew already and building upon it, making
it better. Historian Will Durant writes extensively about how our collective
heritage and knowledge must be handed down generation to generation,
without fail. One break in the chain and we are all savages again. But the
good news is that each generation expands that heritage. He writes that
progress happens "not because we are born healthier, better, or wiser than
infants were in the past, but because we are born to a richer heritage, born
on a higher level of that pedestal which the accumulation of knowledge and
art raises as the ground and support of our being." The dolphins have no
such pedestal.

We used to think that energy was the key component of life, that lightning striking a primordial pool could jump-start it, the way it gave life to

Frankenstein's monster. But now, we understand that it isn't energy but information that matters. Science fiction author Dennis E. Taylor writes about this: "Fire has most of the characteristics of life. It eats, it grows, it reproduces. But fire retains no information. It doesn't learn; it doesn't adapt. The five millionth fire started by lightning will behave just like the first. But the five hundredth bacterial division will not be like the first one, especially if there is environmental pressure. That's DNA. And RNA."

This is correct. It used to be that the only place we stored information was in our DNA. It works okay, but boy, is it slow. It may take millions of years to encode one new piece of information. Brilliant things can be encoded in DNA, such as how the termites build their ventilation systems, or the monarch butterflies complete their migration, and maybe even how *Homo erectus* made a hand ax. But only a very few things can be written there, and only things absolutely needed for immediate survival.

Our breakthrough as a species was that we learned to encode information outside of our DNA. The first place we learned to do this was in our minds, through language. We could externalize what we knew, telling others, and that knowledge could spread like a favorable genetic mutation. Those with that knowledge didn't eat the green berries, a trick that DNA might have taken a thousand generations to learn but now could propagate through our species in a few years. Thus was born Agora. Individual humans became more specialized, learned different things, and, in so doing, our whole became smarter than any individual.

After that, we learned to encode information in writing. This made our virtual genetic code trillions of pages long. Most of it was junk DNA that didn't really help or hurt anything, but not all of it. The learnings of a million lifetimes soon became part of Agora's genome. Thus, Agora has evolved, through natural selection, the ability to make a smartphone. It's as much a part of its biology as making white blood cells is part of ours.

Let's be clear. It's Agora that makes the phone. No human can make a smartphone. Maybe that's not so surprising, but as economist Leonard Read wrote in his 1958 essay, "I, Pencil," no single human even knows how to make a pencil. Think about it: If you were dropped on a desert island, even knowing all that you know, could you make a pencil in a dozen lifetimes? I couldn't. British designer Thomas Thwaites tried to build a toaster from scratch. By scratch, I mean mining the copper, smelting it, and all the rest. He spent years and thousands of dollars on the project, and the "toaster" worked for just a few seconds. But even he didn't really start from scratch; he had the vast storehouse of human knowledge at his disposal, along with the world's financial system, the transportation system, and all the other elements of Durant's pedestal. But Agora can make a smartphone, a pencil, and a toaster almost effortlessly. It was Agora that built the rockets that took us to the moon.

Those four billion years where knowledge could be spread only through DNA got us to the Awakening. Then for fifty thousand years, give or take, we could spread knowledge only through words, orally, in stories and teachings. That got us to five thousand years ago. Then knowledge could spread in written form. This was a huge leap forward because knowledge could persist indefinitely, transcending time and space. You could learn things from people long dead whom you never met. That got us to about 1950. Today knowledge can be stored in infinite quantity and spread in digital form, at zero cost, at the speed of light. Now, Agora's genome is global and changes every second.

Each of these transitional steps has happened far quicker than the prior one. This final change, our ability to encode and spread knowledge digitally, is the first one that transcends biology, so it will advance us even faster. In a hundred years, life on Earth will be as unimaginable to us as our life would have been to Paleolithic man. We will become a single vast intellect, and we will gain mastery over the future. We will do that with artificial intelligence.

Behold the Wonder!

Four out of five animals on the planet are tiny nematode worms, each about as long as a hair is wide. They reached that level of evolutionary success with a brain of just 302 neurons. The industrious honeybee's brain has a million neurons and uses them to build complex, multigenerational societies. An octopus has half a billion neurons, with which it may have achieved consciousness. And then there we are, with brains of a hundred billion neurons—more than the number of stars in the Milky Way—with which we came to rule the planet.

With our vast intellects—and handy opposable thumbs—we built tools to ease our physical burdens. A couple of centuries ago, we started building calculating machines to ease our mental burdens as well. By 1890, we had electric tabulating machines that read punch cards; in the 1930s, we built computers that used relays as switches. By the 1940s, they used vacuum tubes instead. By the middle of the twentieth century, we had created the transistor to replace the tube. Transistors and neurons have a good deal in common: they are roughly the same size, both can function as a single logical circuit, both are powered by electricity, and both can operate in parallel.

Each year those transistors became smaller and cheaper, and used less energy. As they did so, we began grouping ever more of them together to make devices to do increasingly complex tasks. According to Genesis, God created man by molding him out of clay. We in turn fashioned golems from clay as well, that is, the pure silicon from which our transistors are crafted, and every year they advanced in capability. We imagined a day when they could do anything we wanted them to do.

In the seventy-five years since inventing transistors, we've made around five sextillion. That's five followed by twenty-two zeros. It's a number so large that we have no good frame of reference, but perhaps this will do: It is roughly equal to the number of neurons on planet Earth. All of them. The total neurons of every human plus every bird, fish, reptile, all the way down to every little nematode worm. For every fruit fly and mosquito, add another hundred thousand. For every cockroach and beetle, another million. Add all of it together, and that's about how many transistors we have made. We now manufacture them vastly faster than the biome can create neurons. What exactly is it that we are building so frantically? No one is quite sure, but that's not slowing us down any.

We have cast our lot in with our creation and live in a symbiosis with the machines. We are like the Borg now, the half-organic, half-machine aliens of *Star Trek*. The computers need us in order to keep the factories running and the electricity flowing, although they will master those tasks in a few decades. However, we become ever more dependent on them; we literally can't live without them anymore. Imagine a world with no computers. Not just the obvious computers like smartphones and laptops, but the countless billions embedded in everything around us. If they all stop working, then everything stops working. Every tractor, every pipeline, every factory, every-thing. Sure, we all have a few hand tools lying around, perhaps a hoe for working in the garden, but we would even lose our ability to make more of those. The world would go silent, except for the distant sound of the Amish raising a barn.

Having no computers means returning not just to a pre-computer 1950 but to a pre-electric 1890, because the electrical grid is computer controlled, as are the solar panels, windmills, nuclear plants, and refineries. Without electricity, life would screech to a halt, supply chains would break, cities would empty, and somewhere between half and 90 percent of the population would die. Forget the dystopias where the computers take over; the real nightmare scenario is if they stopped working.

We seem fine with this dependency, or at least we are oblivious to the degree to which we have put our fates in the virtual hands of the machines. We love our gadgets too much to really fear them. Computers make life wonderful. They are not just our smartphones and tablets; they power everything around us. Your house probably has about a thousand computer chips scattered throughout it, as does your car.

If we do inhabit a numerical reality, then the computer is our ultimate creation. That's its native habitat, and it is better suited to such a world than we are. As philosopher Marshall McLuhan said decades ago: "The computer is the most extraordinary of man's technological clothing; it's an extension of our central nervous system. Beside it, the wheel is a mere hula-hoop."

Think of how new computers are. Your grandparents, maybe even your parents, are older than the transistor. In less than one human lifetime, we went from computers having no real impact on the world to our being unable to live without them. And it is only going to speed up from here. What they can do, the way they can process data without error at vast speeds, dwarfs our abilities so much as to be incomprehensible. In the 1500s, Ludolph van Ceulen spent his entire professional life calculating pi to thirty-five digits, even having it engraved on his tombstone. Today, a $400 desktop PC from Best Buy running Windows can use a fifty-line program to calculate pi to a hundred million digits in five minutes.

It is easy to see how all of this instills fear in many people. Computers are so alien, so cold, so impersonal. They seem to be everywhere, doing everything, and no one is in charge. It is all just happening.

Upon landing in the New World in 1519, Hernán Cortés ordered the ships burned so his men knew there was no going back. We, too, have burned the ships. We have built a world without an undo button; we cannot go back.

This is all true, but we needn't let it leave us in doubt as to our place in the world. The creation is never greater than the creator. The computer may do some tasks better than we can, but so do the backhoe and the blender. Is the computer fundamentally different from our other tools in a metaphysical sense? I don't think so.

We have talked about emergence—when systems take on characteristics that none of its components have—a few times before. Computers exhibit a kind of emergence. A billion transistors can do things no single transistor can do. But there's nothing inherently mysterious about it. We can marvel that a computer can play *Jeopardy!*, but we can deconstruct every answer forensically and understand exactly why the computer chose it. Computers may exhibit surprising behavior, but nothing that fundamentally can't be understood once we dig in a bit. They are thus deterministic. Philosophically, the computer is no different from a mechanical clock, which, when wound up, runs its program the only way it possibly can.

Are we that way? Aren't we just clocks? No. We exhibit that special kind of emergence where the capabilities of the whole cannot be derived from the parts. It is called strong emergence, and as I mentioned earlier, some people don't believe it exists, that it is an appeal to something almost magical and unscientific. But I think it does exist and it isn't magic. It's quite rare, to be sure, but I think we are a strong emergent phenomenon. I think life itself is such a phenomenon. You can take a cell apart piece by piece and never find life. Yet the cell lives. Life is thus special, arising in a way that we do not understand. There is not a cell in your body that has humor, compassion, pity, remorse, hopes, dreams, or ideas. There is no cell that can fall in love. No part of you has any of those qualities, yet somehow you have all of them, and no one understands how that can be. And then there is consciousness, the fact that you *experience* the world. You can *feel* warmth, which is different from *measuring* temperature. No one knows how matter can experience the universe. By all that we know of machines, there is no way they can experience the world. Thus, humans are not machines.

Therefore, computers are, and will always be, vastly inferior to us. Not just in ability, but in worth as well. In fact, they have no worth, no moral standing. We built them to do one thing, to manipulate ones and zeros at unfathomable speed, and they do that wondrously, but even that is to our credit, not theirs. They have no aspirations to exceed their programming, to better themselves, because they have no self.

But say you don't believe any of that business about strong emergence, that you dismiss it as magical thinking. *Then* should you be worried about computers being our masters or taking over? Again, no. That simply can't happen. We've seen it in enough movies that it looks reasonable, but we are different sorts of things from computers. It looks as if we are the same kind of thing, because as machines do more tasks that only humans used to be able to do, they increasingly seem to be thinking in the same way we do.

But can a computer think? No, that's just a metaphor. Like all metaphors, it contains things that are true intertwined with some that aren't. A computer with a webcam and a microphone doesn't "see" or "hear." Computers don't "decide" or "conclude" or even "understand" anything, but these colloquialisms are a kind of shorthand, which even I use as a convenience throughout this section.

However, I agree that computers certainly *look* as though they think, and some people believe they really do. Mathematician Alan Turing wrote that if we can't tell the difference between a conversation with a computer and with a human, shouldn't we say that the computer thinks, even if it does so completely differently? With all due respect to Turing, I disagree. "Thinking" is a vastly different *kind* of thing from speech. Polymath Danny Hillis made a computer out of Tinkertoys and yarn that plays a perfect game of tic-tac-toe. Imagine a Turing-like test in which the goal was to figure out if you were playing a person or a pile of Tinkertoys in tic-tac-toe. If you couldn't tell, does that mean that the Tinkertoys think? No. If a wax figure at Madame Tussauds is indistinguishable from a person, do we conclude it has crossed a metaphysical threshold of some kind?

Consider the Chinese room thought experiment. It was put forth by American philosopher John Searle. The setup is this: There is a room full of volumes of a special kind of book. In the room is a librarian who speaks no Chinese. Outside the room are Chinese speakers who write down questions in Chinese and slide them under the door. The librarian picks up one of the questions, looks at the first character, and finds the book with that character on the spine. In that book, the librarian looks up the second character, which has a notation to go pull another book and look up the next character, and so on. After the last character has been found, the final book has an instruction to copy a series of characters onto a piece of paper and slide it under the door, which the librarian faithfully does. The Chinese speakers outside the library pick up the message and discover that it is a brilliant answer, both wise and witty. The question Searle poses is, "Does the librarian speak Chinese?" Most people would say no.

The analogy is straightforward: The librarian is a computer and the books are the program. When your GPS is directing you through traffic, the computer doesn't "know" what it's doing. It's just running a program. To say that is the same as thinking is akin to believing a windup walking toy has a will to go across the room.

Computers are wonders. But they are just things, not beings. They also cannot imagine the future; they can only do math. But this is fine, because we have imagination enough for both of us.

AI 101

The first fully transistorized computer came online in 1954. Two years after that, twenty-eight-year-old John McCarthy, an assistant professor at Dartmouth, coined the term "artificial intelligence" (AI) and convened a group of computer scientists for a summer "to find how to make machines use language, form abstractions and concepts, solve kinds of problems now reserved for humans."

I find their naïveté quite charming. Create AI in a summer? In 1956, no less? But upon reflection, their optimism wasn't unreasonable. Simple scientific laws had been found that explained a huge range of observable phenomena in physics, photonics, magnetism, thermodynamics, and quantum mechanics. There was every reason to hope that intelligence would be found to arise from a few simple laws as well. It turned out this was not the case. Intelligence is so complicated that we still don't understand it and can't even agree on a definition of it.

Given that, it is no surprise that there is no formal definition of AI either. There isn't even consensus on the sense in which it is artificial. Is it so described because we made it, as opposed to natural intelligence? Or because it isn't really intelligence, the way artificial flowers aren't really flowers? McCarthy himself regretted coining the term, feeling it set the bar too high for machines.

There are two different things people mean when they use the term AI, and they have nothing in common. But because the distinction isn't well understood, the blanket term AI is used to describe them both, to much confusion, the same way that biweekly can mean both twice a week and every other week. So, when you read a headline about the AI in your spam filter getting more efficient, and another headline about how AI might take over the world and enslave us all, one can't help but begin regarding one's spam filter with a certain distrust. But AI means two unrelated things in those contexts.

The dystopian notion of AI in the "enslave us" sense is something called general AI. It is a program that can do anything a human can do cognitively. It is creative and can teach itself new things. No one knows how to build this or even if it is possible.

I used to host an AI podcast, and on it I asked most of my hundred or so guests, all the leading thinkers in the AI world, if they thought general AI was possible and, if so, when we would get it. Only three guests said it isn't possible. I happen to agree with those three, by the way. Of the vast majority that believe we can build general AI, the range of their estimates for how long from now we will achieve it vary from five to five hundred years. Often, the estimates center around twenty years, but general AI has been twenty years out for seventy years now.

Here's the thing. Those hundred or so AI experts who believe we can make a general AI would all be in 100 percent agreement that we currently don't know how to do it. That's why the time estimates are all over the map. So if no one knows how to make it, why is everyone so confident we can? I would put that question to them as well, and I always got variants of the same answer: "Machines with general intelligence are possible because *humans* are machines with general intelligence."

Those eager to see general AI built are part of a kind of religion, one whose adherents are trying to build a god. It will be, according to them, omniscient. Oh, and benevolent—it will be a kind god, and it will grant the faithful everlasting life, either by the perpetual renewal of their physical bodies or through a kind of metaphysical joining whereby their consciousness is uploaded into the machine, where they will live forever in the harmony of heaven. They believe in the Armageddon, a great apocalypse they call the Singularity, and they believe it is near. And this is all based on a single sacred article of faith: that people are machines.

Others believe the same article of faith but worry that the god we are building will not be benevolent. Perhaps it will be as far above us as we are to ants, and perhaps it will regard us with the same blithe indifference. Or worse, the god will see us as parasites, wasting scarce resources; or as vermin infesting its world; or as a sort of dangerous virus. It doesn't much matter which, for any of them would need exterminating, and the god could do that quite easily.

The reason general AI is such a frightening concept to many is that we have seen it do frightening things before—in movies. So often, in fact, that we do something called "reasoning from fictional evidence." I know I do it. I sometimes catch myself thinking, "That could totally happen. I've seen that before." But then I remember where I saw it before: at a movie theater last summer, and the summer before that. Don't get me wrong, I enjoyed watching the Avengers battle the rogue AI Ultron as much as the next guy, but it doesn't mean that it is real, or even that it could happen any more than the Incredible Hulk could happen. It's just another scary fairy tale, like the kind we first told when we found ourselves in a changing world we didn't fully understand.

Are we machines? That's the big question, the one upon which our collective fate hangs. If we are, then my podcast guests are correct: eventually we will build a mechanical mind, and it will get forever more powerful. If we aren't machines, then humans will remain forever preeminent, at least on this planet. In the speeches I give around the world, I often ask my audiences if they believe *they* are machines, and generally about 15 percent of them raise their hands. It is an amazing disconnect: 97 percent of the AI experts I interview believe they are machines, while only 15 percent of the public does.

The question is a simple one of fact—we either are or aren't. But the *belief* that we are is, I believe, fundamentally toxic because it undermines the basis of human rights, that there is something so extraordinary and transcendent and unique about being a human that instantly gives you— regardless of your merit, wealth, or ability—special status. However, if we are just machines, then killing a human takes on the same moral characteristics of powering down a laptop. What's the difference?

Few groups are working on trying to create a general AI. I can count on two hands the number of organizations I know of that are seriously trying to build it. The vast majority of the money is spent on the second kind of AI. The good news is that in our quest to see and master the future, we don't need general AI at all, so we can set aside that philosophical debate and focus solely on the other kind of AI, narrow AI, a tool we already use extensively. It is an entirely different kind of technology.

A narrow AI is a computer program that can perform a single cognitive task. Examples include spam filters, GPS navigation, and thermostats that learn your preferences for temperature. Our approach to narrow AI, which I will just call AI from this point on, has gone through three distinct phases. The first was to try to model a particular aspect of the world in a computer. For instance, if you wanted an AI to play tic-tac-toe, you would write an AI that embodied that game, coding the strategy the way a human player would think of it. This is a straightforward approach, but it works on only the simplest tasks. After that came what are known as expert systems. With those, you would find the best tic-tac-toe player in the world and say, "For each of the nine possible first moves, how would you respond?" Then, for each response, again ask the expert what they would do. This approach is handy in a few places such as industry, in which the problems are well defined and the choices limited. Both of these approaches, you may notice, are capped by our level of human understanding, unlike the third one. This third approach is the one we will explore, because it is how we use AI to predict the future right now. It is called machine learning. A machine-learning approach to a tic-tac-toe program would be to teach the computer the rules of the game, but no strategy, and then have the computer play itself millions of times to teach itself the best way to play.

Machine learning is how we do the vast majority of AI today. To teach a computer how to identify a photo of cats, you give it thousands of photos of cats as well as thousands of not-cats, then start feeding it new photos, asking if it is a cat and telling it if it is right or wrong. Over time, the computer learns what a cat looks like.

You've no doubt seen this—and even been a party to it. Years ago, the test used by websites to make sure you weren't a robot required you to type some text from an image that was fuzzy or missing pixels. You were volunteer labor helping a corporation train its AI optical character recognition program. Then, a few years later, the test changed. You were shown photographs of numbers on curbs, and by deciphering those you were donating your time to help train some company's AI to read street addresses. Now there is a third test where you are asked to find all the stop signs or crosswalks in a photo. When you do that, you are kindly working for free to train some business's self-driving cars.

Why are humans needed for all of this? Because the ability of humans to pattern-match vastly exceeds that of a computer. If you spot a friend at a distance, on a foggy morning, walking away from you, you still may recognize them from some small peculiarity in their gait or just a glimpse of the back of their head. That's amazing. Also, you can be shown a new object—something you have never seen before—and then can instantly pick it out of photos even if it is mostly obscured, underwater, or a different color. A child can be shown a single photo or even a drawing of a cat and then pick out cats all day long. If the child happens to see a Manx cat, they may say, "Look, a kitty with no tail," even though they have never been told such a thing exists, because they have such an innate pattern-matching ability that they can perceive some subtle aspect of "catness" that we all can recognize but can't explain.

No one has any idea why we are so good at pattern-matching or how we do it so well. I could show you a drawing of an alien—a single drawing—and you could find it in photos even if it was looking away or lying down. You could spot it in a watercolor painting or stylized as a black-and-white film noir character, or even if it was rendered as a baby or very aged. All from a single drawing of a creature that doesn't even exist.

Part of the reason we can do this is that we are really good at something else computers score a zero on: transfer learning. That's where we take knowledge from one sphere and apply it to another. The trick we have somehow mastered is knowing what can transfer and what cannot. Here's a thought experiment: Imagine a trout just caught by a fisherman from a stream one minute ago. Now imagine an identical trout in a jar of formaldehyde in a laboratory. Which of the following do they have in common? Weight? Length? Smell? Temperature? Color? I'm guessing this test wasn't too hard for you, even though it is doubtful you have firsthand experience with trout in formaldehyde. But how did you know the answer? If you can figure that out, let me know and we'll split the Nobel money. Computers can't do transfer learning like that at all. Teaching an AI to identify cats doesn't get you any closer to an AI that identifies dogs, so we are left training AIs on one thing at a time.

Our storytelling nature would probably never have evolved if we couldn't do these things so well. Stories work because we recognize the patterns in them and can use transfer learning to apply them to other things. "The Boy Who Cried Wolf" isn't all that interesting unless we can relate it to similar situations, none of which involve boys, crying, or wolves. If anything, our pattern-matching abilities are a little too good. We see animal shapes in clouds and faces on toast. In his book *Explore/Create*, Richard Garriott writes about the creation of his role-playing game *Ultima Online*. One insignificant, almost throwaway feature was that if you were tired of epic questing, you could grab a fishing pole and drop a line in the water, which was stocked with "generic-looking fish" that you would catch 50 percent of the time. It was a simple coin toss whether you caught a fish or not. That's all. However, Garriott writes that players thought they saw patterns in their successes, saying that they began "believing that they had better results when fishing two or three yards farther offshore than when they cast close to the riverbank. Fishing at night, some people were convinced, was more productive than fishing in the afternoon. People created their own mythology."

This is how superstitions arise. If a baseball player doesn't change his socks once and achieves an upset victory, you will be able to smell him coming all next season.

Computers pattern-match quite differently from us. We're smart, but they aren't. They rely solely on brute force. They take all those photos of cats and not-cats and break them down into tiny clumps of pixels. They look at all the two-pixel combinations and three-pixel ones, and so forth. Every pixel is assigned a number corresponding to its color, and the billions of clusters across all the photos of cats are, to the computer, tiny photos of cats. Feed in a new photo and the computer will chop it up into bits and compare them to its massive database of other bits. Any two pixels may or may not be a cat, and the computer's best guess on that question is probably only 50.001 percent accurate, but the fact that it can do billions of calculations a second means that it gets higher levels of confidence by working harder, not smarter. Both the law of large numbers and the central limit theorem are the backbone of modern machine learning.

The machine doesn't know what a cat looks like, or even what a cat is, but it grows ever better at spotting them based on pure mathematical models. The computer never says something is or isn't a cat; it just returns a value between 0 and 1 of how likely it is that it is a cat. It involves the same basic calculations as dice rolls and coin tosses. All machine learning is thus probabilistic. Pascal and Fermat would have no problem understanding the math. Heck, they invented it. What would be beyond their comprehension is the scale of it all. But in all fairness, it is beyond our comprehension as well.

When the computer accurately says something is a cat, we naturally assume it comes to this conclusion as we do, by knowing what a cat looks like. But it doesn't. If you were to print out the calculations that the computer went through to figure it out, it would just be a bunch of numbers that happened to be able to identify cats. This fundamental difference between how humans and machines pattern-match is why "explainability" in AI is so hard. "Why did the AI think this was a cat?" can't really be answered in English. A lack of explainability isn't a big deal when identifying cats, but for loan applications it is, especially to the person who was just denied a mortgage because the computer said they were likely to default but could give no understandable reason why it came to this conclusion.

I belabor this a bit just to point out that a computer isn't looking at a photo thinking, "I'm pretty sure that's a cat. Look at those pointy ears." In fact, the computer isn't thinking at all. It's just comparing a bunch of ones and zeros relating to colors of clusters of pixels. It is nothing more than a fancy windup clock.

The bottlenecks with AI right now are in cleaning up data and then telling the AI when it is right and wrong. The dream is for something called an unsupervised learner, where you could point at the internet and it would just figure things out without a human saying, "Right, wrong, wrong, right, right," a million times. We are a long way from having this, however, and no one is quite sure how to build it—or even whether it can be built.

Digital Mirror

Getting clean, labeled data with which to train our AIs isn't a difficult task per se; it's just a terribly slow and time-consuming one. Because of this, all the places where we *could* apply AI are waiting their turn to get their data all sorted out. If all innovation in AI stopped today, we would have decades of work just to do the things we already know how to do with AI.

There are applications of AI where we aren't the bottleneck, where the data is unambiguous and clean—because that is data recorded by sensors. Sensors are devices designed to gather specific types of information. They have a wider range of capabilities than our biological sensors and can operate in more hostile environments. They are governed by their own form of Moore's law, so they're always getting better while their prices constantly fall. To date, we have deployed about a trillion of them. Before sensors, computers were sealed up in boxes with no access to the outside world, relying on humans to spoon-feed them data. When sensors are attached to machines, they can see and hear. They can gather their own data without our involvement at all.

Sensors open up a vast range of new applications for computers. For example, a typical smartphone has a dozen or so sensors that connect it to the outside world. One of these is a location sensor based on GPS. Many phones constantly broadcast their location, completely anonymously. When enough people do this, the traffic flow for the whole country can be recorded in real time. In effect, the computers are replicating our analog world in their digital memory, operating as a sort of digital mirror. So while the traffic flows around the actual streets, it concurrently flows around a virtual one as well. Of course, it is all just a bunch of ones and zeros, but that is what a digital reflection of our analog world looks like.

With this digital mirror, the computers can optimally direct traffic for the whole country. If they see that there are a hundred phones at a certain spot on the interstate that are moving at just three miles per hour, they can reasonably assume there has been a collision there. If they further see that the traffic on the adjacent access road is unaffected, they can route the traffic off the interstate onto the access road. But not all the traffic, just enough to balance things out. The machines can do this without our ever touching the data. This application of computers and sensors is already being utilized and is tremendously beneficial. Every day that the system runs, more data is collected to further train the algorithms, so they always get better. I've noticed that the program I use on my phone to route through traffic usually gets my arrival time accurate to within a minute or two. Talk about seeing the future.

With the price of sensors falling, in some cases, to a fraction of a cent, we will be able to build systems like this for more and more parts of our lives, not just traffic. We won't just computerize everything but sensorize it as well. We've already connected fifty billion smart devices to the internet, and we are on a path to fifty trillion.

As we explored earlier, we used to be able to store data only in DNA. Then with writing, data expanded exponentially, creating our virtual genome. Movable type and cheap paper meant the genome grew more, and then with the advent of the digital age, it grew exponentially again. Now you often see statistics like, "We create more data every ___ days than we did since the dawn of time." Those stats are pretty meaningless since it really is apples to oranges, but they do get at something big, that Agora's genome is expanding rapidly. Where does this ultimately lead?

Imagine for a moment if we produced so many sensors that we were able to log everything—that is, if we created a digital reflection of not just our traffic but our entire world. What are we likely to record? Everything. Start with every word you say, every place you go. Every person you meet, and everything they say to you. Everything your eye tracks to, along with your physiological response to it. Imagine if every breath you take were logged, and every beat of your heart. Every object you own would be enlivened with sensors. Your clothes, your jewelry, your furniture, all the way down to your toothbrush. Every pot and pan, every appliance. Every utensil would be loaded with sensors that would log the nutrients of every bite you take. When you go to a store, everything you handle but don't buy will be logged; when you dine out, exactly what you ordered and how much of it you ate will be as well. Every interaction with every person, every flower you stop to smell, how hard you clap at a concert. Every word you type, every dollar you spend.

This will happen, not because Big Brother will force it on us, but because we will demand it. If a spoon can keep me from getting salmonella, I want one. If a toothbrush can tell me I am coming down with a cold, I want one of those, too.

Perhaps this is your idea of a dystopia. If so, hold on to that thought a minute. For now, just consider the good that it could do.

Recall the five shortcomings of humans that limit our ability to see the future, to be masters of our own fate. We are bad at math, poor at reasoning, poor at collecting data, limited in the amount of data we can handle, and unable to comprehend the complexity of the world we are in. Computers with sensors are perfect at math and can be flawless in their reasoning. They can collect an unlimited amount of data and analyze it as well. And with enough processing power, they can figure out the manner in which everything is connected to everything else.

What sorts of connections are these? No one had any idea that iodine deficiency was so terrible and yet so prevalent in the US. But after iodized salt was introduced in 1924 under the promise that it would eliminate goiter, the IQ of the entire country went up 3.5 points. In states with high levels of iodine deficiency, it went up fifteen points. In the United States, the stereotype of Southerners as lethargic and dim-witted was often true among the poor who went barefoot and picked up hookworms from the soil. It was epidemic in rural areas in the South, and a 1926 study on the topic said that a person infected with hookworm seemed to be "living in another, entirely separate world, and is only remotely in contact with the everyday world about him." Why did this just pertain to the South? According to the 1940 US census, at that time 94 percent of people in Massachusetts had flush toilets, while only 19 percent had them in Mississippi. Digging the holes in outhouses just a bit deeper drastically reduced the incidence of hookworm. Solving this one problem raised the IQ of the entire region. A similar blight on the South was niacin deficiency due to a largely corn-based diet. Fortifying cornmeal with niacin further increased the overall IQ and solved a host of other health problems. Removing lead from paint increased IQ across the nation, while removing it from gasoline had an even larger impact. Since cities had the highest density of automobiles, they had the highest density of environmental lead as well. Lead from cars is believed to have caused the rise in urban crime throughout the 1960s and 1970s. After its elimination, crime fell and, once again, IQs rose.

A digital mirror of our world would have easily uncovered all of this. The iodine, the hookworms, the niacin, the lead—everything is there in the data, just screaming out at us. But our world is a cacophony of data all screaming at us, and things only randomly rise above the din, such as when some smokers on the antidepressant Wellbutrin reported that their cravings for cigarettes declined, and it was found that the drug is a powerful smoking cessation aid. It's now sold as such under the trade name Zyban.

History is full of these sorts of random discoveries. People in Russia and Finland used to keep a brown frog in their milk to prevent it from spoiling. Only later did we discover that the frog's secretions are antibacterial. In World War II, German soldiers in Africa reported that the locals cured dysentery by consuming fresh, warm camel dung, which we now know contains a powerful antibiotic.

One wonders how any of this was discovered. Who was the first person to put a brown frog in their milk and just happen to notice the milk lasted longer? Or who was fond of eating fresh camel feces and one day happened to notice their dysentery was cured? Who knows what we haven't yet discovered? Perhaps yodeling while doing Russian squat kicks will add ten years to your life. And conversely, just imagine what we might be doing today that is stupefying us all. Perhaps in a few years we will wake up to a headline such as "Apples: Nature's Silent Killers."

No one can write down everything they know. And even if they did, who could read it all? Who knows how many times someone discovered the camel dung cure before it stuck? Before, we could only save the really good stuff. So Plato's writing has survived, but his aunt's bunion cure has been lost to the ages. Now, imagine if collectively nothing were ever forgotten. Imagine if the life experience of every person who lives from this moment forward were preserved forever, and that data were used to improve the lives of everyone to come. Just think about how life would be different if this technology had been invented a thousand years ago, and today we lived in a world where our choices could be informed by each of the choices, both good and bad, of the billions of people who came before us.

Having access to this technology, our descendants will marvel that we made any progress at all. To them, our lives will look as though we were drunken sailors on shore leave, staggering through life making capricious decisions based on faulty reasoning and anecdotal data. Only every now and then did one of us eat some camel dung or put a frog in our milk and learn something new.

We need the computers and sensors to better our lives, to allow everyone access to the wisdom of the ages. We can't collect all the data ourselves and try to make sense of it without machines because our brains aren't up to the task. Imagine if every little decision everyone has made over the past thousand years along with its outcome had been recorded on index cards and stored in a gargantuan facility somewhere. Remember that giant warehouse at the end of the first *Indiana Jones* movie where they ended up storing the Ark of the Covenant? That's where index cards AA through AC are housed. Imagine five thousand more of those to store all that data. What could we do with it? Nothing useful.

Computers can do only one thing: manipulate ones and zeros in memory. But they can do that at breathtaking speeds with perfect accuracy. Our challenge is getting all that data into the digital mirror, to copy our analog lives in their digital brains. Cheap sensors and computers will do this for us, with prices that fall every year and capabilities that increase.

Coupling massive processing power with sensors will create a species-level brain and memory. Instead of being billions of separate people with siloed knowledge, we will become billions of people who share a single vast intellect. Comparisons to *The Matrix* are easy to make but are not really apropos. We aren't talking about a world without human agency but with enhanced agency, information-based agency. Making decisions informed by data is immeasurably better. Even if someone ignores the suggestion of the digital mirror, they are richer for knowing it. Imagine having an AI that could not only tell you what you should do but would allow you to insert your own values into the decision process. In fact, the system would learn your values from your actions, and the suggestions it gives you would be different from those it would give everyone else, as they should be. If knowledge is power, such a system is by definition the ultimate in empowerment. Every person on the planet could effectively be smarter and wiser than anyone who has ever lived.

I believe this system to be a good thing, and I regard it to be inevitable. Besides, what could possibly go wrong?

The Fears

The fears we have around AI are numerous. We are worried that computers will compete with us economically, making more and more humans unemployable. I don't agree with this; I believe instead that the machines will generate more demand for workers by creating higher-paying jobs. There will be a real shortage of human workers very soon, and this will force up wages. It's a great time to be a human.

Why do I think this? Imagine if you could go back in time a quarter century, to the nascent days of the consumer web. Say you showed someone the most popular web browser of the day, Netscape Navigator, and said: "In twenty-five years, billions of people will use this every day of their lives. What do you think it will do to jobs?" If they were a particularly forward-thinking person, they might say, "This will be bad for jobs. It will hurt the travel agents, the stockbrokers, the Yellow Pages people, the shopping malls, and the newspapers." And they would have been right about all of that. But they would not have said, "There will be Google, Facebook, Twitter, Netflix, Amazon, eBay, Etsy, Uber, Airbnb, DoorDash, Spotify, Shopify, Zillow, and literally a million other companies. There will be social media consultants, web designers, influencers, streamers, and, well, thousands of jobs that won't even have names yet."

That's the challenge, right? From our vantage point, we can only see what technology will destroy, and we lack the imagination to see what it will create.

I've long been interested in understanding the half-life of a job. I think it is fifty years. I think every half century half of all jobs vanish. So why do we usually have nearly full employment? Because there are an infinite number of new jobs. Jobs are created out of thin air, out of ideas. You can instantly create a job by taking something, adding labor or technology to it, and thereby making it worth more. The difference between what it was worth and what you can make it worth is a wage.

Because technology enables everyone to be more productive, everyone is able to create more value and earn more. That may sound too idealistic, but imagine if you lived a hundred years ago in a small town. What were your employment prospects? Probably very few. The sewer pipe factory that was the lifeblood of the town? But what about today? With the technology we have, we are empowered to add more value to more places. With technology, the world is your market, communication is essentially free, you have access to an unfathomable range of tools, and you can work for companies in any of a hundred countries. The possibilities really are endless. And this trend is only going to increase. We now have more ideas on what we can do with technology than we have people to do it. That means we really do have a shortage of people, and that will get ever more pronounced as the gap grows wider between what technology could, in theory, do and what we have enough people to do.

Another fear is that AI will concentrate wealth by exponentially increasing the productivity of those who can deploy the technology at scale—that is, the already rich. There is truth to this. But is inequality *in and of itself* bad? What if technology doubles the wealth of the poor but triples the wealth of the rich? That's more inequality, but more wealth for everyone. Is that worse than the status quo? Perhaps, if you regard it primarily as an issue of equity rather than of economics, though if so, that is a matter better addressed through taxation and governance than by limiting technology.

Some believe that AI isn't increasing the wealth of the poor at all but just the opposite: wages have stagnated, and the absolute wealth of the poor is shrinking. Is this true? The answer is complicated. Unquestionably, in the US, there is a large group of people who think so, who feel left out of the prosperity of the modern age and, burdened with student debt and low-wage jobs, have abandoned the twin hopes of starting a family and owning a home, the classic embodiment of the American Dream. This is troubling. Homeownership became a national priority after World War I because it was seen as a way to prevent communism from taking hold in the US. As one real estate organization touted, "socialism and communism do not take root in the ranks of those who have their feet firmly embedded in the soil of America through homeownership." It is certainly true that no one will fight to defend a system in which they have no vested interest or that they think is rigged against them in the first place.

The challenge is that while technology raises human productivity, the financial benefit of those increases flows to whoever owns the technology. If you are a lawyer and a new piece of software lets you write a will in half the time, then you get to pocket the increased profit from that. If you are a checkout clerk who sells their time by the hour, then a new piece of technology that doubles the speed at which you can check people out does nothing for you; that increase goes into the pocket of whoever owns and deploys the technology. It's not a conspiracy; it's just the way things shake out under our system. If we don't like it, well, that's why we form governments in the first place, to, in the words of the US Constitution, "promote the general welfare." In other words, through legislation we can craft the sort of country we want where the benefits of increased productivity are shared in ways that we regard as equitable.

However, the question of the ultimate impact of technology on wages isn't so clear-cut, in part because the way we measure prosperity is now out of date. It wasn't long ago that only the richest people in the world could hear music while they cooked, eat ice cream in the summer, travel without walking, or any of the other thousand things our modern world provides to most of us. And no one, not even the rich, could fly through the air at the speed of sound, watch events as they happened in other parts of the world, or talk to anyone anywhere as if they were in the room with them. Where does any of that get measured in our calculations of wealth? This isn't me saying, "Let them eat tech," because if you can't make rent, you take little comfort that your internet is fast. Rather it's a critique of using GDP as the main measure of national success, since that metric counts *all production* as positive, regardless of what is produced. Robert Kennedy made this criticism a half century ago, pointing out that GDP "counts special locks for our doors and the jails for the people who break them. It counts the destruction of the redwood and the loss of our natural wonder in chaotic sprawl. It counts napalm and counts nuclear warheads and armored cars for the police to fight the riots in our cities." He says that it tells "everything about America except why we are proud that we are Americans." So we may be measuring the wrong things and should instead figure out a way to quantify societal happiness and health. Don't get me wrong: the picture by those standards might be even worse, but at least we would be looking at the right metrics.

There are many more worries about AI. They include algorithm bias, autonomous weapons, cybercrime, lack of transparency, digital warfare, and, perhaps most of all but least mentioned, how it makes our civilization more brittle, more prone to complete systemic failure. The worries about AI are so numerous that they could (and do) fill multiple books. Instead of tackling them all, let's narrow our focus and concentrate on the worries about the Digital Mirror in particular, about the world of cheap sensors and processors.

The first worry is that this kind of information could lead to despotism so pervasive that you won't even be free in the private musings of your mind. I'm not worried about this. I am terrified of it. The privacy and freedom that we have had throughout history has largely been protected by the fact that there are so many of us, and no state can watch everyone, right? No government can listen to every phone call, read every letter, keep everyone under constant surveillance. Well, now they can, and pretty affordably. The disturbing truth is that the AI tools we are building for noble purposes such as looking for early signs of cancer are easy to adapt for nefarious uses such as looking for dissidents in society. It isn't computer overlords any of us have to fear; that's just a science fiction plot. It is human ones who, if they had such a system, could use it to control people forever—as George Orwell wrote in *1984*, "We know that no one ever seizes power with the intention of relinquishing it."

The Chinese government has developed a far-reaching surveillance system built with AI that takes the video from 650 million cameras scattered throughout the country and, using facial recognition software, is able to keep tabs on everyone. Officials make no apologies for it. Quite the opposite—it is a source of national pride. They call it social credit, and the goal is to keep a digital dossier on everyone to encourage better behavior. Each person starts out with a slug of points, and their actions either add points to that number or take them away. There is no secret about any of this; it's no conspiracy theory. The government publishes a list of things you can do to lose and gain points, so everyone knows how the score is being kept. After all, it's hard to discourage bad behavior if no one knows exactly what they aren't supposed to be doing. Jaywalking makes you lose points, but winning a national prize of some kind gets points added. Through this mechanism, people are sorted into two groups, what F. Scott Fitzgerald called the "beautiful and the damned." The goal of the system, as its advocates like to say, is to "allow the trustworthy to roam everywhere under heaven, while making it hard for the discredited to take a single step." *China Daily*, an official publication of the Chinese Communist Party, often carries headlines celebrating the system, such as "China Boasts World's Largest Social Credit System." One article explains that for those whose scores dip too low, "penalties include curbs on taking flights, trains, employment and educational opportunities."

With such a system, a despotic state could criminalize anything—being left-handed, for instance—and punish it any way they wished. Though it's not yet possible with today's technology, the ultimate system using the Digital Mirror could read your biometrics at a distance and notice when you saw a photo of some dissident on a magazine cover at a newsstand that your pupils dilated, indicating an unspoken affinity for them. Reeducation camp for you—or worse.

The Digital Mirror doesn't inherently *have* to be oppressive. The data needn't be centralized, it could be anonymized, and improper use of it could be penalized or even criminalized. How can we avoid the dystopian version? It won't be easy. It has long been said that the price of liberty is eternal vigilance, so we must be ever diligent in our watching of the watchers. Not building the Digital Mirror isn't really an option, though. It's already being built. We seem to want it quite badly. A tweet of actor Keith Lowell Jensen sums it up: "What Orwell failed to predict is that we'd buy the cameras ourselves, and that our biggest fear would be that nobody was watching."

There is a second fear around a system like the Digital Mirror. With enough data, it will be able to make confident recommendations about many aspects of your life. It will see your future, or the possible futures of various different choices you could make. It will tell you where you should eat lunch, what you should order, and when you should return to work. You won't *have* to listen to it, but you will just find life is better when you do. If you take a metal detector to the beach, and it starts beeping, indicating where a gold doubloon is buried, are you going to say to it, "You are not the boss of me. I dig where I like!"? Sure, you can dig anywhere you want on the beach, and maybe you'll get lucky, but the whole point of the metal detector is that it can tell you exactly where to dig.

The machine will recommend where you should go to college, what you should study, which job offer you should take, whom you should marry, and what you should name your firstborn. It will invest your money, suggest your next car, choose a location for your next vacation, and weigh in on every decision where data could be helpful. In such a world, do we become mentally duller? Probably. Our Paleolithic forefathers were probably always on the alert, keenly tuned into their environment. As we have self-domesticated, we have grown more like house cats and less like tigers. Are we okay with this? Maybe *we* aren't, but our grandchildren who will live in that world wouldn't want it any other way.

There is a limit to what the Digital Mirror can show. Chaos always kicks in. It is impossible—and I use that word in its most literal sense—to build a sensor sensitive enough to collect the data needed to overcome chaos. The Planck scale, named for the nineteenth-century physicist, is the most basic granularity of the universe, like the pixels that make up reality. No sensor can measure down there, or even within many orders of magnitude. The Digital Mirror won't be omniscient, or anywhere near flawless. It will simply be better than us, and thus we will come to depend upon it.

It will also never predict something it has never seen before, because it has no imagination. It will only be able to inform decisions based on data in the past and will work only in areas where the future is like the past.

There is a final worry about the Digital Mirror. The story of John Henry is a folktale about a railroad laborer whose job was to make the small holes in rock into which explosives were inserted to clear the way for the track. One day a steam-powered drill was invented, and John Henry challenged it to a contest. He won it, but only with such an exertion that he died, "hammer in his hand."

This tale is often interpreted as a manifestation of the economic competition that machines presented to human labor in the nineteenth century. But it isn't that at all. John Henry wasn't thinking about money; he could have gotten a better and easier job operating the steam drill. No, he felt that the machine undermined his personhood. In one telling, he told the steam drill inventor: "A man ain't nothin' but a man. Before I let your steam drill beat me, I'll die with a hammer in my hand."

In a *Jeopardy!* tournament, an IBM computer named Watson famously beat the best human player, Ken Jennings. Jennings made a comment that echoed that of John Henry. Having seen a chart on which the IBM engineers graphed the increasing proficiency of Watson in a hypothetical matchup with Jennings, he remarked, "And I realized, this is it. This is what it looks like when the future comes for you. It's not the Terminator's gun sight; it's a little line coming closer and closer to the thing you can do, the only thing that makes you special."

So that final question is whether we are somehow less human in the world of the Digital Mirror, where the future is clear to the machine, and we depend on it for guidance? Are we missing something that John Henry got—that when machines do our work, it doesn't ennoble the machine but lessen the human? Today, most of us aren't bothered by power tools the way John Henry was. But is there anything we wouldn't turn over to the machines if they were better at it than us? Is there any situation where we would want to hold back and say, "This far and no further? If the computer does this, then we are all less for it"? What if computers could be better judges than humans? Better therapists? Better doctors? Better politicians? Better friends? Better spouses? Better parents? Is there any point where we say to the AI in our best Gandalf voice, "You shall not pass!"?

As a society we don't know the answer to this question. And while the possibility of machines being potentially better parents or spouses is a ways off, many of the most-hoped-for uses of AI are to provide companionship for the elderly and assist with day care.

In 1966, AI pioneer Joseph Weizenbaum created a simple AI called ELIZA, which was what we today would call a natural language chatbot. One of the modules created for ELIZA was a sort of therapist designed to parrot back whatever you said as a question. If you typed, "I am feeling sad," it would reply, "Why are you feeling sad?" If you answered, "Because of my mother," it would say, "What about your mother is making you sad?" and so on.

Although the system was pretty simple—after all, this was 1966—many people became attached to the program, pouring their hearts out to it, convinced it was doing something more than just repeating variants of what they said in question form. Seeing this, Weizenbaum turned against his creation and later went on to write a book called *Computer Power and Human Reason*, in which he argued that any job that required empathy should never be performed by a computer. This includes everything from eldercare to customer service. He maintained that interacting with machines that emulated human empathy would have a corrosive effect on us, making us feel devalued and isolated.

Weizenbaum also made the distinction between deciding and choosing, maintaining that deciding is a computational task suitable for machines, such as deciding the shortest way from point A to point B. Choosing involves values and should never be delegated to the machine. A human should be the one to choose to drive their car into a tree to avoid hitting a child.

This distinction is acknowledged by Will Smith's character, Detective Spooner, in the movie *I, Robot* when he explains why he hates robots. He tells the story of his car colliding with another car, becoming intertwined, and plunging off a bridge into the water. He was alone in his car, but the other car had a driver who was now dead and a twelve-year-old girl, who, like him, was trapped. Water was pouring into both cars, soon to kill them both. But nearby there was a robot that witnessed all of this and jumped into the water with just enough time to try to save one of the people. But which one? It computed the odds of success for each and made a decision.

Detective Spooner: I was the logical choice. It calculated that I had a 45 percent chance of survival. Sarah only had an 11 percent chance. That was somebody's baby. Eleven percent is more than enough. A human being would've known that. Robots [indicating his heart] . . . nothing here, just lights and clockwork.

Weizenbaum feared that we would be tempted in the future to delegate our humanity away and that this tendency stemmed from an "atrophy of the human spirit that comes from thinking of ourselves as computers." Recall that 97 percent of guests on my AI podcast thought of themselves as machines.

Our best instincts urge us to show compassion and kindness to all things that can feel pain, that have a "self." Computers don't have a self. But we already build computers that talk in human voices and have human names. This is problematic, I think. I once brought home a robot dog for a week to test it out for an article I was writing. The puppy had big eyes and a cute "yip," and it would walk around exploring my house. My kids would interact with it, absentmindedly petting it as one might a real puppy.

As a species, we've had to fight long and hard to become civilized and to develop empathy for other living creatures. But imagine if you got an AI to help take care of Grandpa in his waning years. Say it looked something like a human, had a name—let's call it Alfred—and a human voice. Alfred wouldn't be a general intelligence, just a really souped-up version of ELIZA. It would listen to Grandpa's stories, again and again, tell ones of its own, assist him in getting around the house, remind him to take his medicine, and so on.

If Alfred slipped in the kitchen and broke, would we be fine pitching it into the landfill? Probably not. People want broken Roombas fixed, not replaced, so you can only imagine that Grandpa would want *his* Alfred back. But what if we did throw it away? Can you imagine a landfill with a bunch of broken Alfreds strewn about? Things that looked like you and me, lying there, abandoned. And then, if Grandpa slipped in the bathtub and broke his hip, are we now more willing to throw him away? Weizenbaum feared we would be.

The Far Future

O ur play in three acts began when some human somewhere learned to think in language. That person passed that gene to the next generation, where it spread throughout the group. We learned to vocalize our thoughts and share them with each other. This was the birth of humanity. Anatomically, *Homo sapiens* had been around a long time, but fully modern humans were something entirely new, no longer just animals.

Our brains also allowed us to recall specific events in the past at will. This was new, too. The flip side of that coin was that we could imagine specific events in the future. We could consider possible futures and choose one to try to actualize. We became the unquestioned rulers of this planet. The curtains closed on Act I.

Act II opened in the Renaissance, when our growing prosperity had produced men of science who had means enough to tackle the weighty questions of the world full-time. Their forebears had looked to the night sky and wondered, but then went to bed because they had to get up early to plow the fields. This new generation had leisure enough to avoid the fields, so they woke up thinking about these problems as well. They asked all the same questions as before, but they set about to answer them in a different way, through observation, hypothesis, and experimentation.

One of the questions they tackled was related to causes. "Why do things happen the way they do?" That was followed by "How can you predict what will happen next?" They found order in nature in a place no one had expected to find it: randomness. They came to have a different conception of the future, a probabilistic one. After that, the next three hundred years were spent riffing on that scientific truth. The sophistication of their methods advanced to a point where they needed mechanical help with cognition to advance further. So they built an electronic neuron. The curtains close on Act II.

Act III is interesting. It is based not on a new discovery about the world but on a new invention. We still use the same techniques we've used for the past century, but we've built a device to employ those techniques at a scale hitherto unimagined. Using that massive processing power, we have begun to amalgamate our collective knowledge into a global brain and have enabled it to grow over time on its own. This is such a radical break with the past that it is hard to see where it will take us, except in outline.

I have said a few times that this book is about seeing the future, but today our ambition has expanded past that. No longer content to just *see* the future so we can influence it, we now want to control it. In the movie *Lawrence of Arabia*, there is a recurring theme of events being written, that is, fated to happen. But at one point, Omar Sharif's character, Sherif Ali, says in awe of T. E. Lawrence, "Truly, for some men nothing is written unless *they* write it." We want to write our own fate.

Looking at our entire play holistically, we see certain trends that have held true over the past fifty thousand years, despite occasional setbacks. What are they? They are trends toward longer life, more self-rule, more liberty, more self-determination, more education, more literacy, and an increased standard of living. In addition, there has been progressively less hunger, less poverty, less disease, less crime, less violence, and less war. We no longer torture people for popular entertainment, and in most places, we've outlawed cruelty to animals as well.

There is still much to do. We've created the idea of human rights but haven't yet applied them to everyone. We've eliminated legal slavery, but the illegal variety still exists. Women have achieved legal equality in many places, but in others they are still regarded as chattel. Child labor has been outlawed in most countries, but in other countries children are regularly worked to death. Our challenge is to continue to right the wrongs around us until someday we find ourselves in a world that finally lives up to our noblest aspirations, a world in which we have been able to, in the words of Aeschylus, "tame the savageness of man and make gentle the life of this world."

This book is about seeing the future, and this is the future I see for us all. It is not a dream, nor a wish, nor a hope—it's a confident prediction.

EPILOGUE

S ometimes things don't make sense until the end of the story.
    The question I have tried to answer in the past eighty thousand or so words is this: How did we spend eighty thousand or so generations using those never-changing Acheulean hand axes, but then somehow got from there to the iPhone in just 2,500 generations? Something dramatic changed in us that opened our eyes, enabling us to mentally transcend time and view its flow from the past to the present to the future as an infinitude of stories full of meaning and significance, not simply a series of events, a cosmic domino rally that has been going on for 13.8 billion years. Each of our individual lives is not a sterile narration of a tedious sequence of about 2.5 billion seconds.

Earlier, we looked at the twenty purposes of stories. But there is one more, a twenty-first one, that I held back. It's the special one: **Stories give life meaning.** One story of our existence is that humans are bags of self-contained chemical reactions and electrical impulses that careen through space bumping into other bags of chemical reactions until one day we fizzle out and our meaningless lives dissolve into nothingness and are quickly forgotten. It's the bleakest view of life possible because on a cosmic scale it renders our existence absolutely pointless. It is the logical conclusion of determinism, of pure rationalism. It is a worldview that can be summed up in just two words: "Shit happens."

But there is another story, the one most people really do believe, that they *know* to be true even if they cannot rationally defend it. It is that our lives *do* matter, that everyone has dignity and worth, that we are all here for a purpose. The truest truths in life are not equations because, as poet Muriel Rukeyser wrote, "The universe is made of stories, not of atoms."

This isn't mere poetic license. Science teaches us a wonderful fact, that each of us is made up of the debris of stars that burned out eons ago. We are, as Carl Sagan said, "star stuff." But as amazing as that is, it misses our true essence. We are, each of us, a story, which means that the events of our lives are connected not just by causality, but by meaning, which must be imputed to it by the storyteller. Thus, we can answer the age-old question, "What is the meaning of life?" by answering a slightly different version of it:

"Who is telling your story?"

ACKNOWLEDGMENTS

I want to thank my family—my wife, Sharon, and our four children—for their constant support and encouragement throughout the entire process of writing this book. Also, I want to thank Alexa Stevenson, my editor extraordinaire, for her amazing insights and feedback; Glenn Yeffeth, my publisher—fourth time's a charm; and of course my agent and good friend Scott Hoffman, who is always there every step of the journey with sage advice and insightful suggestions. Also, Robert Brooker and Ellis Oglesby, who offered invaluable observations on the manuscript; Nancy Watkins, who carefully read through the whole book twice, suggesting hundreds of edits along the way; Gregory Newton Brown for doing a thorough read-through with insightful line edits; Scott Calamar for his careful and contemplative copy edits; Rachel Massaro for help fact-checking; the University of Texas's Jason Abrevaya, who did Zoom calls with me on weekends to help with the dice section; and my eye doctor, Dr. Tom Walters, who posed an interesting question to me during an exam that got the ball rolling on the whole thing. In addition: John Erickson, Doug Hohulin, Pablos Holman, Jason Horton, Brett Hurt, Steve Baughman Jensen, Steve Lanier, John Matthews, and Stephen Wolfram. Finally: Christina Berry, John Connally, Pamela B. Erwin, Dave Grimme, Mike Lemper, Orenthia D. Mason, Patricia Meyer, Shari Nelson, Jo Northcutt, and Kevin Stambaugh.

INDEX

ABOUT THE AUTHOR

Byron Reese is a longtime serial entrepreneur who has started a half dozen companies over the past thirty years. He grew up on an East Texas farm, graduated from Rice University, married his college girlfriend, moved to the West Coast, and relocated from there in the 1990s to Austin, where he and his wife raised and homeschooled their four children. Along the way, he fell in love with technology and began writing about its significance to the story of humanity. His first book was a defense of techno-optimism called *Infinite Progress*. He followed that with a philosophy book about AI and robots called *The Fourth Age*. In 2020, he and his coauthor Scott Hoffman released *Wasted*, a popular science book about the nature of waste. This book, *Stories, Dice, and Rocks That Think*, is about how humans learned to see into the future and how that radically altered our trajectory as a species. Today, Byron is CEO of an AI company he started and is working on his next book. He and his wife still reside in the Austin area.